PRACTICAL GUIDE TO WORK STUDY
[REVISED EDITION]

KERWIN MATHEW

PRACTICAL GUIDE TO WORK STUDY [REVISED EDITION]

PREFACE

This book serves as an easy-to-understand guide to engineers, production supervisors and managers, laymen and students on the subject of work study.

In any organization, be it the office where filing work and typing activities are carried out, or the factory where production of goods takes place, there is inevitably inefficiency, such as unnecessary movements resulting in fatigue in the person performing the task and hence inducing even greater inefficiency, and longer time taken to perform a task than is necessary. This is a big headache for management, resulting in loss of earnings and in some cases loss of business and even closure of company. The heavy responsibility of work study or methods analysis falls squarely on the shoulders of the work study officer or the industrial engineer.

In most organizations today, e.g., in shipyards and manufacturing companies, work study officers and industrial engineers are employed in droves. These "efficiency experts" are indeed worth every cent they are paid. They help the organization to reduce wastage, cut down costs and increase productivity and profits. Thus, in an essential way, they help to ensure the survival of the organization.

For students taking up a course or examinations on industrial engineering, this book will certainly give them a good, easy grasp of the subject of work study. For production management staff and industrial engineers, it will serve as a handy reference text. For the layman, it will be a "teach yourself" book on work study. To add a practical touch to the book, and to challenge the reader, several case studies have been included, which if properly carried out, should provide him a good feel for the subject.

For anyone who is at all concerned about efficiency, this book will demonstrate to him what efficiency really is.

Kerwin Mathew, Ph.D., PE, CMfgT, CPM

CONTENTS

1 A FEW WORDS ABOUT WORK STUDY

Work study or methods analysis concerns the development of more efficient work methods. More efficient work methods draw the firm closer to its goal of maximizing profit. Thus, in work study, it is essential to study the way in which work is being carried out, with a view towards developing a procedure which if adopted would serve to increase the profits of the firm.

There are no hard and fast rules to work study. It should however be emphasized that the best time to develop a good method is when preparations are made to perform a task for the first time. The same principles would of course be involved as are involved in the analysis of an existing method. However, there is no need for describing the existing method as none exists. But every proposal for an original method should be analyzed and treated as though it were an existing method.

Finally, it should be emphasized that though the human element is important to method improvement, other aspects such as plant layout and materials handling should also be looked into.

2 INTRODUCTION TO WORK STUDY

INTRODUCTION
Work study can be defined as the study of methods of working in order to discover ways of making more economical use of manpower, materials, machinery and money. It is divided into two elements as follows:-

WORK STUDY comprises of:

[i] **METHOD STUDY** (to improve methods of production, leading to better use of:
Manpower, Material, Equipment, Plant)
[ii] **WORK MEASUREMENT** (to evaluate human effectiveness, leading to improvement in:
Staffing, Planning & Control, Incentive Schemes)

DEFINITION OF WORK STUDY
Work study is a generic term for the techniques, such as method study and work measurement, which are utilized in the evaluation of human work in all its contexts and which systematically lead to the investigation of all the factors that affect the efficiency and economy of the work situation under review, so that there will be improvement.

Methods analysis concerns finding the best way of performing the job.

Work measurement concerns finding the length of time the job takes to complete, through time study.

For instance, if one were trying to find a way of improving the method of washing commercial vehicles, one would first use method study to find the most efficient work system, questioning every aspect of the current way of performing the job.

After having found a better method, one could use work measurement to find out how many man-hours are required in washing one vehicle. With this information, the manpower requirements could be established and a financial incentive scheme could be implemented to encourage the staff to maintain a good work-rate throughout the day.

Questions for Review

1. Describe the relationship between methods analysis and work measurement.

3 WORK MEASUREMENT

INTRODUCTION
A lot can be written about work measurement. In this chapter, an overview of work measurement, which is only one aspect of work study, will be given.

WHAT WORK MEASUREMENT IS
Work measurement is the process of measuring or forecasting the rate of output of an existing or newly designed operation and determining how much time is taken up for the various productive and nonproductive activities of an operation, process or job, as well as determining the standard times which represent the allowable time for the performance of the work. Work measurement, being a generic term, concerns all the techniques of time measurement of the work systems. The results of work measurement are used as an analytical tool for the evaluation of work methods, determining the standard time values for given tasks, cost analysis and comparisons, and developing standard time data systems.

LABOR STANDARD
The actual work standard may differ greatly from the scientifically established industrial engineering standard. The informal organization with its own system of authority, work standards, leaders and communication network may also have some impact. The informal organization should not be ignored by the operations managers who should attempt to influence it to communicate its work standards, at the same time attempting to influence the acceptance of formal standards by the informal work group.

A labor standard stipulates what an average worker performing under average job conditions is expected to achieve. The following important questions should be considered when setting a labor standard:-

[i]   How is an "average" worker determined?
[ii] What is the appropriate performance aspect to be evaluated?
[iii] What measurement scale should be utilized?

After the answers to these questions are obtained, work measurement techniques can be used to establish labor time standards.

People differ not only in physical characteristics such as height, strength and arm span but in working speed as well. It is necessary to find an "average worker" in order to establish a labor standard. But how do we find the "average worker"? A typical worker if chosen may not be typical in every aspect. It may be best to observe several workers to observe several workers and gauge their average performance. The costs of sampling have to be traded off against the costs of inaccurate standards. The greater the number of workers sampled and studied in depth the higher will be the total cost of establishing a standard. For instance, if we evaluate each of seven workers for one hour instead of each of three workers for an hour the cost of the evaluation (sampling cost) becomes more than double. However, the advantage is that when more and more workers are sampled and studied the closer the performance standard should be to the true "average" performance. Inaccurate standards come with costs, e.g., inefficiencies are tolerated and product costs are distorted. This affects all the uses of standards. An accurate standard of course can never be guaranteed. But increasing the number of workers studied can reduce the total costs of inaccuracy. A range of reasonably low total costs can be achieved through trading off the costs of sample size and the costs of inaccuracy.

After the average performance rates are determined, the performance standard has to be established. Should the standard be set at the average of total performances for the group or at a level at which almost all in the group can be expected to achieve? For example, should the standard be set at 21.75 units per hour, the average, or at 12 units per hour, a level that 95 per cent of the workers can be expected to achieve? Either is acceptable in fact. It may be felt that adopting a minimum standard, the second choice, encourages poor performance and it is better to have about one-half the workers seeking but not attaining 100 per cent of the standard, i.e., setting the standard at the mean performance (21.75 units per hour). Others may feel that standards should be attainable by 90 to 95 per cent of the work-force. Either approach can be effectively used.

SETTING PERFORMANCE STANDARDS

With regards to the setting of work standards, quantity is generally considered to be the primary aspect to be gauged while quality is the secondary standard. In manufacturing quantity is usually measured as pieces per time period while in the service industry it is usually measured in service units per time period. For instance, for a timber sawing operation the set standard performance might be 1,200 pieces sawed per hour and for a bank telling operation the set standard performance might be 20 customers served per hour. The quality standard set is often based on the percentage of defectives, i.e., the number of defective units divided by total units and multiplied by 100. The quality standard for the sawing operation might be set, e.g., at one per cent allowable defective units and the telling operation at 0.05 per cent error in coin-counting. The most important points to consider in determining the various aspects of performance are as follows:-

[i] The aspect should be specified before setting the standard.
[ii] The standard and subsequent actual performance aspect should both be measurable.

A work measurement scale wherein the normal performance is scaled at 100 per cent can also be used. Such a scale is shown below:-

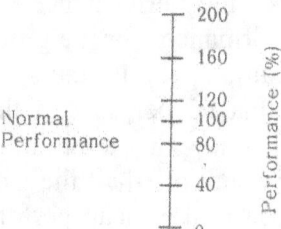

If the performance, e.g., is 30 per cent above normal, the production of the worker is 130 per cent of the normal scale.

ACCURACY
The experienced raters can evidently set a more accurate standard than the inexperienced raters. The standards set by the experienced raters generally have lower variability than the standards set using only historical data, though there are also errors. Although setting standards is not a finely developed scientific procedure and there are likely to be some errors, raters should be utilized for work measurement.

WORK MEASUREMENT TECHNIQUES
There are six basic ways of setting a time/work standard, which are as follows:-

[i]   Disregarding formal work measurement.
[ii] Using the work sampling approach.
[iii] Using the predetermined time study approach.
[iv] Using the direct time study approach.
[v]   Using the historical data approach.
[vi] Combining [ii], [iii], [iv] and [v] above.

DISREGARDING FORMAL WORK MEASUREMENT
Formal labor standards are not set at all for many jobs in many organizations, especially in the labor-intensive service industries - a fair day's work for a fair day's pay does not happen in actuality. This results in ineffective administration or poor management. A worker may be blamed for below par performance and inefficiency though this criticism is not justified. Poor labor efficiency is likely to happen if workers were not given specific, understandable goals. Usually because a work/time standard has not been established some informal standard becomes established by default. Formal work measurement should thus not be neglected.

USING THE WORK SAMPLING APPROACH
Of the six basic techniques of setting a time/work standard, work sampling, which was pioneered in the 1930's, is the most recently developed. While in many of the other techniques gauging with a stop-watch is carried out, in work sampling measuring with a stop-watch is not resorted to. Work sampling relies on simple random sampling techniques derived from statistical sampling theory instead. The following describes how work sampling is carried out:-

[i]   Determine the conditions for the definition "working" and the conditions for the definition "not working". "not working" refers to all activities which are not specifically defined as "working".

[ii] View the activity at selected intervals and record whether the person is working or not.

[iii] Compute the proportion of time a worker is engaged in work (P) using the following formula:-

P = x/n = No. of observations wherein working occurred/Total no. of observations

The manager can estimate the proportion of time a worker is engaged in work with the above computation. This proportion can then be utilized as a performance standard.

Work sampling can also be used for establishing production standards, with a procedure that is similar to the one used in direct time studies. Normal times can be calculated with the formula below:-

Normal time = Total study time x Percentage of time employee observed worked x
            Performance rating factor/No. of pieces produced

Standard times can be computed with the following formula:-

Standard time = Normal time/1 - Allowance fraction

Work sampling is especially suitable for service sector jobs, e.g., jobs in government, health care, insurance companies, banking and libraries. The accuracy of these techniques depends much on the sample size. Though a larger sample size incurs higher cost, it leads to increased accuracy, precision and reliability being important. For setting the sample size for the various reliability levels, a basic statistics book can be of help.

Work sampling can be extended to include output standards by incorporating a concept known as rating or leveling performance. After a job has been studied, the work study analyst has to determine whether the worker's performance was average, above average or below average. If it is decided that the worker's performance was average no adjustment will be made. If it is decided that the worker's performance was above average (gauged as units/time period) the worker's rate is multiplied by a factor less than one. If it is decided that his performance was below average the rate is adjusted to average by multiplying the observed performance rate by a factor greater than one.

The accuracy of this performance leveling depends greatly on the skill and judgment of the work study officer or industrial engineer. A lack of objectivity may be there and the results may vary from study to study. Moreover, only a few workers may be subjected to the study. Also, "working" is a broad concept which is not easy to define with precision. There are however some evident advantages with work sampling. It is an economical way to gauge work performance, simple and easily adapted to the service industries and the jobs involving indirect labor. In fact, if used with discretion work sampling is a helpful work measurement technique.

USING THE PREDETERMINED TIME STUDY APPROACH

The predetermined time study approach is helpful in setting standards for jobs which are not presently being performed but are being planned. Besides the direct time study method, this method can also be used for existing jobs. It involves stop-watch time study and time study from films. Historical data of tens of thousands of people performing such basic motions as grasping, lifting, reaching, standing and stepping are collected. Industrial engineers break these motions down into elemental actual times which are averaged into predetermined standards and published in table form. The following is the procedure for establishing a predetermined time standard:-

[i]   Observe the job which is yet to be established or think it through. It is best to use representative materials, a typical machine and an average worker performing the job correctly when observing the job.

[ii] Record each job element, do not worry about elemental times and thoroughly document all the motions performed by the worker.

[iii] Get a table of predetermined times for various elements and record the motion units for the various elements; motion units are expressed in some basic scale (a Therblig scale often being used) which corresponds to time units.

[iv] Sum up the total motion units for all elements.

[v]   Provide an allowance for personal time, fatigue and delays in motion units.

[vi] Sum up the performance motion units and allowance units for a standard job and convert these motion units to actual time in minutes or hours; this total time is the predetermined time standard.

An example of a methods time measurement chart is given below. The motions of the right and left hands are shown in this chart, codes are also provided and the TMU (Time Measurement Unit) motion units are also shown. The chart shows how a predetermined time standard is established. The method used here is known as Methods Time Measurement (MTM) and is a widely accepted predetermined time study approach. The technique allows a job to be observed, breaking the job down into movements which are studied in depth and which have a predetermined average time. For clarity, the MTM chart shown below is split into several blocks. The error allowance (placing an error card in the error box) and the allowance for personal needs, unavoidable delays and fatigue are near the bottom of the chart. The time allowance for personal needs, unavoidable delays and fatigue represents a standard industrial engineering allowance. For the job under consideration, a 15 per cent allowance, which is a widely used allowance, is adopted. The total MTM time per cycle, 397.9 TMU (as is indicated in the chart below), is directly converted to 0.23874 minute per cycle, i.e., 4.188 units per minute, or, 251 units per hour, if one TMU equals 0.00001 hour.

Right Hand	Code	TMU	Code	Left Hand
		14.2	R12D	Reach to cards
		3.5	G1B	Grasp a card
		10.6	AP2	Apply pressure to separate
		3.5	T455	Turn card
		13.4	M12B	Move to focus eyes
Transfer card from other hand	G3	5.6		
Subtotal		50.8		Subtotal
Multiplied by 6		304.8		Multiplied by 6
		5.6	G3	Transfer cards from other hand
		13.4	M12B	Move to final box
		4.0	D1E	Disengage cards
		15.0	WP(1)	Walk to start again
Subtotal		342.8		Subtotal
Error allowance		3.2		Error allowance
Subtotal		346.0		Subtotal
		51.9		15% personal fatigue, and delay allowance
Total TMU		397.9		Total TMU

METHODS TIME MEASUREMENT CHART FOR ESTABLISHING THE QUANTITY STANDARD

The major advantage of predetermined time studies is that non-representative worker reactions to direct time studies are eliminated. As the standard is set away from the work-place in a logical, systematic manner, workers do not slow their work-pace or get nervous. With this method, as compared to direct time studies, disruptions on the shop floor are less severe as the workers do not experience anxiety. The basic disadvantage of this technique, which is encountered early in its use, is that if some job elements are not recorded or are recorded improperly future timing would not be accurate. If job elements could not be properly identified and tabulated they should be evaluated with the direct time study approach.

USING THE DIRECT TIME STUDY APPROACH

This method, often called time study, stopwatch study or "clocking the job", is most widely used for setting work standards in manufacturing. The industrial engineer using this method simply studies a job, with stopwatch and clipboard in hand.

There are basically six steps involved in this method, which are as follows:-

[i]   Observe the jobs which are being timed. This method depends on direct observation and is thus limited to jobs which already exist. The job selected for the study should be standardized, in terms of materials and equipment, and the operator performing the job should be representative of all operators.

[ii] Select a job cycle, identifying the work elements which constitute a complete cycle. Determine how many cycles are to be timed with a stopwatch.

[iii] Time the job for all cycles. As workers behave in varying ways when their performances are being recorded, the common reactions being nervousness, resentment and slowing the work-pace, minimize these effects through repeated study, study across several workers and standing by one worker while studying a job somewhere nearby, e.g., in another department.

[iv] Complete the normal time based on the cycle times.

[v]   Determine the allowances for delays, fatigue and personal time.

[vi] Set the performance standard, i.e., standard time, as the sum of observed normal time and determined allowances - the sum of steps (iv) and (v) above, which can be expressed by the following formula:-

Standard time = Normal time/1 - Allowance fraction

where

Normal time = Average cycle time x Rating factor

Average cycle time = $\sum$ Time recorded to perform an element/Number of cycles observed

Allowance fraction = Fraction of time for unavoidable work delays, fatigue and personal needs

$0 \leq$ Allowance fraction ≤ 1.0

When timing jobs, industrial engineers frequently use a rating factor. A worker is hence judged as 85 per cent normal, 90 per cent normal or some other rating, depending on the industrial engineer's perception of what is normal, which is subjective. A sample could be chosen because the industrial engineer intuitively judged that the sample was of a size large enough to provide a reasonably accurate estimate of the average time at a reasonable cost, there being an accuracy/cost tradeoff in direct time study.

USING THE HISTORICAL DATA APPROACH
In this method, past performance is assumed to represent normal performance. Some managers use past performance as their principal guide in setting standards, in the absence of other formal techniques.

The historical data approach is inexpensive, simple, quick and perhaps better than not establishing a work standard at all. The main disadvantage of this method is that the past might not be representative of what an average worker could achieve under average working conditions. Also, some of the historical data may reflect the performances of extraordinarily capable or incapable workers or unusual working conditions. The historical approach however may misrepresent average performance. Management should therefore intuitively adjust past performance data upward or downward prior to applying them as a standard. Many organizations have utilized the method successfully to achieve profitability, survival and growth over long periods of time despite its shortcomings.

COMBINING WORK MEASUREMENT TECHNIQUES
Work measurement techniques are employed in combination, as cross-checks, in practice. It is a common practice to observe a job, jot down in detail all the job elements and establish a predetermined time standard. After this, the history of performance on this job or similar jobs should be checked to verify that the predetermined standard is reasonable. The job by elements and in total could be time studied to provide a further check. It should be borne in mind that no one work measurement technique is entirely reliable. Setting a standard requires great skill and is not easy. Whenever possible, there should be cross-checks.

WORK MEASUREMENT FOR WHITE COLLAR WORKERS
The same measurement techniques which are employed in the service industries are evidently appropriate for white-collar jobs since the latter are typically labor-intensive and minimally automated. A combination of work sampling and historical data may be the most suitable. Predetermined time study can be useful (where it can be used) for the more routine white-collar jobs.

USE OF WORK MEASUREMENT

A comprehensive survey on whether U.S. and Canadian industries utilized work measurement and, if they did, their reasons for having done so, had been carried out in 1976. Nearly 1,500 responses had been obtained in this survey, wherein 89 per cent of the respondents reported that they were using work measurement, which contrasts sharply from the 1956 survey in which only 71 per cent of the 785 respondents reported that they had used work measurement. In the 1976 survey 53 per cent of the respondents had been found to have used work measurement to gauge employee performance as compared to only 20 per cent in the 1956 survey. The 1976 survey also showed that work measurement had been utilized for production scheduling (55 per cent), setting wage incentives (59 per cent) and estimating and costing (89 per cent). All this indicates that work measurement is by no means "outdated" or "dead" in the industries and is quite useful.

Questions for Review
1. Define work measurement.
2. Describe the relationship between work measurement and methods analysis. Which typically follows the other and why?
3. Explain how individual job standards differ from departmental and plant standards. Give an example of each from an organization of your choice.
4. Provide two uses of time (labor) standards. Explain the way the time standard could help a department within an organization of your choice for the two uses you have chosen.
5. Explain the predetermined time study approach.
6. In establishing a standard, why would combinations of work measurement approaches be a good strategy?
7. Explain how a standard for a group of seven draftsmen in a large architectural firm could be set.
8. What is the importance of production/operations standards?

4 NATURE AND DEFINITION OF WORK STUDY

INTRODUCTION
Work study can be described as the study of the ways of doing things or the critical examination of existing and proposed methods of work with the aim of improving them. Its objective is the elimination of unnecessary work and waste, e.g., waste of time, materials or capital equipment, which can be achieved by improving the overall process, the work environment, the efficiency of the use of human resources and materials, the working procedures, and, the layout and design of the building, equipment or workplace.

DEFINITION
Work study is the systematic recording and initial examination of existing and proposed methods of doing work, with the aim of developing and applying easier and more effective work methods and reducing cost.

OBJECTIVES OF WORK STUDY
The overall objective of work study is to achieve greater efficiency and higher productivity. Work study contributes to greater efficiency through the following:-

[i]   The improvement of procedures and processes.
[ii]   The improvement of workplace and factory layouts.
[iii] Improvement in the utilization of manpower, materials and machines.
[iv]   Standardization of work methods.
[v]   The improvement of the work environment.
[vi]   Improvement in material handling.
[vii] Design improvement.
[viii] The improvement of standards of safety.

METHOD ADJUSTMENT
When there are difficulties or dangers encountered in work methods it is a quite common practice to devise safeguards or take measures to get round the difficulties, which people commonly do in their everyday life. This is known as method adjustment, which concerns only those aspects of the work method which encounter problems. Though adjustment of the work method may occasionally be necessary, it does not represent an improvement of the whole method. Planned method improvement relies on a system of analysis that is used to evaluate the whole method and can be expected to bring the following results:-

[i]   Better utilization of manpower.
[ii] Improved utilization of equipment and machines.

[iii] More effective utilization of materials.

The following is an outline of the system of analysis:-

METHOD IMPROVEMENT MASTER SHEET
We first ask the important question: Can the whole job be eliminated? If the job can be eliminated there is no need to improve it.

For improving the work methods in a section the following can be implemented:-

[i]   Decide which job to start on.
[ii] Collect all the facts about the existing work method.
[iii] Examine these facts.
[iv] Develop a better work method.
[v]   Take measures to get the better work method working.
[vi] Keep an eye on things to ensure that the new, improved work method is maintained.

The method improvement master sheet can be regarded as a practical plan to make better use of manpower, materials and machines. It is indeed important to determine whether the whole job can be eliminated. In method improvement, we should select a job which we can quickly:-

[i]   Make safer.
[ii] Make easier.
[iii] Reduce excessive or unnecessary movements.
[iv] Get rid of bottlenecks.

We should:-

[i]   Obtain the cooperation of the worker.
[ii] Observe the work being done.
[iii] Chart the present work method.
[vi] Record snags and difficulties.

We should also examine the work method by first challenging the "DO" activities and then the remaining activities.

We should look into the following:-

[i]   What is achieved? Is it necessary? Why is it necessary? What other achievements would be better?
[ii] Where is it done? Why is it done there? What other place would be better?
[iii] When is it done? Why is it done at that time? What other time would be better?
[iv] Who does it? Why is it done by that person? What other person would be better?
[v]   Consider: Safety, Quality, Equipment, Materials, Layout, Design.

Note all ideas.

We should develop the work method by carrying out the following:-

[i]   Review ideas and take note of trends.
[ii] Eliminate. Simplify. Combine. Rearrange.
[iii] Make new work method SAFE.

We should install the work method through implementing the following:-

[i]   Consider the best time for introducing the work method.
[ii] Convince all concerned about the work method.
[iii] Train the workers using the work method.

We should maintain the work method by:-

[i]   Checking the progress frequently.
[ii] Watching results.

We should look for opportunities for further improvement.

BASIC PROCEDURE
There should be a definite and ordered sequence of analysis in evaluating any problem, which may be summarized as follows:-

[1] SELECT   -   the work that is to be studied.
[2] RECORD   -   all the relevant facts about the present work method by direct observation.
[3] EXAMINE   -   these facts critically in an orderly way, utilizing the techniques most suited to the purpose.
[4] DEVELOP   -   the most practical, effective and economic work method, bearing all

		contingent circumstances in mind.
[5] INSTALL	-	this work method as a standard practice.
[6] MAINTAIN	-	this standard practice through regular routine checks.

CHECKLIST ON INTRODUCING CHANGES

[1] Have the reasons for changes been explained?
[2] Have the benefits derived from a particular change been explained?
[3] Has everybody been informed about who will be affected?
[4] Have they been informed soon enough?
[5] Is there any opportunity for them to make comments?
[6] Are their suggestions given attention?
[7] If a suggestion is rejected are the reasons provided?
[8] Are significant comments brought up to management?
[9] Would the change be better introduced gradually?
[10] Has how the change will affect the workers' self-respect been considered?
[11] How will the workers' security be affected by the change?
[12] How will the workers' sense of belonging to a team be affected by the change?
[13] Can a way in which people can save face in accepting the change be provided?
[14] What support and assistance can be provided to the workers to help them during the change?
[15] Will there be patience and tolerance for those who will now find the going tougher?
[16] Are the ultimate objectives firmly kept in mind?
[17] Are you prepared to make adjustments in your plans for achieving these objectives?

Questions for Review
1. State the six steps in the method improvement master sheet?
2. Explain method adjustment.

5 SURVEY AND SELECTION OF JOB FOR WORK STUDY

INTRODUCTION
Certain factors should be borne in mind when considering whether a particular job should be investigated. These factors are as follows:-

[1] At all stages, economic considerations will be important. If the economic importance of the job is small or if it is one which is not expected to continue for long, it is evidently a waste of time to commence or to continue a long investigation. The following questions should always first be asked: Will it be worthwhile to begin a method study of this job? Will continuing this study pay?

The following may justify an investigation:-

[a] Bottlenecks that are causing delay to other production operations.
[b] Operations which involve a lot of labor and equipment or movements of materials over long distances between shops.
[c] Operations involving repetitive work utilizing a great deal of manpower and liable to continue for a long time.

[2] The technical aspects should be considered. It is most important to ensure that adequate technical expertise is available with which to carry out the method study.

The following are examples:-

[a] Pertaining to the loading of unfired ware into kilns in a pottery, a change in work method may result in increased plant and labor productivity but there may be technical reasons why this change should not be made, e.g., a specialist in ceramics may advise against the change due to technical reasons.
[b] A machine tool may cause a production bottleneck due to it running at a speed below that at which high-speed cutting tools will operate effectively. The machine tool expert should look into this problem; he can consider the following: Can the machine tool be speeded up? Is the machine tool robust enough to take the faster cut?

[3] Mental and emotional reactions to investigation and changes of work method, which are most difficult to predict, have to be anticipated. With experience pertaining to the local personnel and conditions, these difficulties could be reduced. Workers representatives, union officials and the workers themselves should be informed of the reasons for and objectives of the method study. However

promising the study of a particular job might be from the economic point of view, it should be abandoned if it appeared to lead to ill-feeling or unrest. If other jobs were performed well and could be seen by everyone to benefit the people working on them opinions would change and, in time, it would possible to return to the original choice.

If the first subjects selected for the method study are the ones which are unpopular such as dirty jobs or jobs requiring the lifting of heavy objects, the method study would be more willingly accepted by the workers. The method study would be seen to be reducing the effort and fatigue of the workers and would be welcomed by them if these jobs could be improved, with the unpleasant aspects removed from them.

SELECTING A JOB FOR WORK STUDY
We can adopt the following procedure in selecting a job for work study:-

[1] Select a task within a job that will permit significant and quick improvement.
[2] Select a simple job for which one is directly responsible and which is already being considered in a large improvement scheme.
[3] Check with management whether the job will be a useful one to study. Choose a job with the greatest potential for improvement first though practically all methods are subject to improvement at some time or other.
[4] Select any job in which there is danger, a bottleneck, excessive movement, or which is strenuous or difficult.

FIELD OF CHOICE
It will be helpful to have a standardized list of points to cover when selecting a job for method study, which prevents factors being overlooked and allows the suitability of different jobs to be easily compared. Below is a sample list which is quite comprehensive but lists should be adapted to individual requirements:-

[1] Product and operation.
[2] Person who proposes the investigation.
[3] Reason for the proposal.
[4] Suggested limits of the investigation.
[5] Details of the job:

  [a] How much is* produced or handled per week? (*For bulk materials gauged in pounds, tons, kilograms, feet, meters, et al.)

[b] What approximate percentage is this of the total produced or handled in the shop or plant?
[c] How long will the job take?
[d] What is the quantity required in future?
[e] How many workers are employed on the job directly, and, indirectly?
[f] How many workers are there in each grade and on each rate of pay?
[g] What is the average output per worker (or team of workers) per day?
[h] What is the daily output compared with the output over a shorter period of time (e.g., one hour)?
[i] What is the mode of payment (e.g., piece-rate, time-rate, team work, premium bonus, et al.)?
[j] What is the daily output of:

  [i] the best operator?
  [ii] the worst operator?

[k] When were the production standards set?
[l] Are there any specially unpleasant or injurious features in the job? Is the job unpopular with:

  [i] workers?
  [ii] supervisors?

[6] Equipment:

  [a] What is the approximate cost of plant and equipment?
  [b] What is the current machine utilization index?

[7] Layout:

  [a] Is the existing space allowed for the job sufficient?
  [b] Is extra space available?
  [c] Is it necessary to reduce the space already occupied?

[8] Product:
  [a] Are there frequent design changes which cause modifications?
  [b] Can the product be modified for easier manufacture?
  [c] What is the quality required?
  [d] How and when is the product inspected?

[9] What increase in productivity or savings can accrue from the method improvement?:

[a] Through the "work content" of the product or process being reduced.
[b] Through better use of machines.
[c] Through better utilization of labor.

(Figures may be provided in man-hours, machine-hours or money or as a percentage.)

Questions for Review

1. How would you select a job (with which you are familiar) for work study?

6 RECORDING: TECHNIQUES OF RECORDING

INTRODUCTION

After selecting the job to be studied, it is necessary to record all the facts relating to the existing work method. The success of the whole procedure is dependent on how the facts are recorded as the facts will provide the basis for both the critical examination and the development of the improved work method. It is important that the facts are recorded clearly and concisely.

THE NEED FOR RECORDS

It is essential to have on record all the necessary facts of the existing work method so that the activities selected for investigation may be visualized in their entirety with a view to improving them by subsequent critical analysis. A record is necessary if a "before" and "after" comparison is to be made to evaluate the effectiveness of the investigation and the subsequent installation of the new work method.

The recording of facts is usually done by writing them down but this method is unsuitable for the recording of the complicated processes which are so common in today's industry, especially so when an exact record is required of every minute detail of an operation or process. This is a time-consuming process as even a very simple job which may take a few minutes to perform would probably need to be recorded in several pages of closely written prose which would need careful study before it could be grasped in detail.

Other techniques of recording have been developed to overcome this difficulty, so that information may be recorded precisely and in standard form in order to be readily understood by all those involved.

Before attempting to improve the existing work method it is important to find out precisely what is currently being done. This recording is carried out as follows:-

[a] Get the cooperation of the person who is performing the job.
[b] Observe the work being done. Look out for bad work habits or outdated procedures. Spend some time with the worker, if possible, helping him to carry out the job, this being a good way of experiencing the difficulties first hand.

NOTE SNAGS AND DIFFICULTIES

Get a broad view of the job first by observing the different tasks in the total job. For instance, a caterer who occasionally serves executive buffet lunches in companies goes through the following procedure in preparing the food:-

[i]   Identify the food items needed.
[ii]   Draw packaged and tinned food from the store.
[iii]   Purchase fresh food from the farms.
[iv]   Clear the working space in the common room annex.
[v]   Prepare the food.
[vi]   Get ready the table.
[vii]   Serve the food.
[viii] Collect the dishes.
[ix]   Wash up and clean up.

The snags and difficulties associated with each of the above tasks can, e.g., be recorded as follows:-

[i] Identify food items which are required. Guests sometimes have special dietary needs.
[ii] Serve food
    - Food should be carried through on trays.
    - Food is occasionally dropped due to overloading or poor design of the trays.
    - Coffee cups should be washed though nobody likes doing it.

Charting can be used to carry the recording of the present work method a stage further.

RECORDING TECHNIQUES
Charts and diagrams are the most commonly used recording techniques. See below:-

[A] CHARTS indicating process SEQUENCE

    Outline Process Chart
    Flow Process Chart - Man Type
    Flow Process Chart - Material Type
    Two-Handed Process Chart

[B] CHARTS using a TIME SCALE

    Multiple Activity Chart

[C] DIAGRAMS indicating MOVEMENT

    Flow Diagram

String Diagram

PROCESS CHARTS

When one is trying to understand the process in order to arrive at a better work method, a long written explanation of how a job is performed can be rather difficult to grasp. To overcome this difficulty process charts are utilized in work study to provide a clearer picture.

The following are the two main types of charts used:-

[i] Outline Process Chart
[ii] Flow Process Chart

OUTLINE PROCESS CHARTS

Unless the job to be studied is relatively simple, an overall view of the process to decide on the need for detailed recording and subsequent analysis is initially required. The outline process chart is particularly useful for obtaining this overall picture.

DESCRIPTION

The outline process chart is the simplest type of process chart. It is a diagrammatic illustration of the two main activities, operations and inspections, in a process. To describe these two activities, standard symbols are utilized. These symbols are linked together to form the chart which is defined as: "A graphic representation of the points at which men or materials are introduced into the process and of the sequence of all operations and inspections associated with the process, except those associated with materials handling."

Units of time, quantity and distance may be added as notes to indicate when and where work activities take place, in order to make the chart more useful.

SYMBOLS

The symbols used for the two main activities are as follows;-

Operation O
[i] The above symbol represents an operation.
[ii] It is defined as follows:

"An operation occurs when an object is intentionally changed in any of its physical or chemical characteristics, is assembled or disassembled from another object, or is arranged or prepared for another operation, transportation, inspection or storage. An operation also occurs when information is given or received, or when planning or calculating takes place."

[iii] The main result is production or accomplishment. According to the nature of the work an operation takes many forms. Depending on the amount of details which the chart is intended to record the magnitude of the work represented by the symbol can vary from chart to chart. The symbol should consistently represent operations of equal magnitude on one chart.

Inspection □
[i]   The above symbol represents an inspection activity.
[ii] It is defined thus:

"An inspection occurs when an object(s) is examined for identification, or verified for quality or quantity in any of its characteristics."

[iii] The main result is verification. The following are typical activities which are classified as inspections:

[a] Verification of quantity - number, measure, weight, et al.
[b] Verification of quality - test to a standard, grading and color, et al.

Operations should not be confused with inspections. An operation produces, accomplishes or changes something while an inspection only verifies that this has been correctly carried out. An example is presented below:

① Prepare Carton
② Load Machine
③ Load Acessories
☐ Inspect Cartons
④ Seal Cartons

CONSTRUCTION OF AN OUTLINE PROCESS CHART

Below is an example of an outline process chart for assembling a switch rotor, which is a simple operation with activities divided into operations and inspections and the appropriate symbols allotted accordingly:-

EXAMPLE OF AN OUTLINE PROCESS CHART:
ASSEMBLING A SWITCH ROTOR

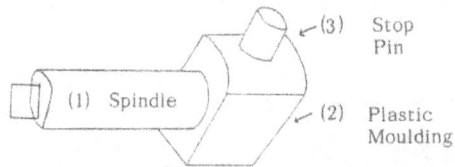

The above switch rotor consists of the following parts:

[a] A spindle (1).
[b] A plastic moulding (2).
[c] A stop pin (3).

When making an operation process chart insert a vertical line down the right-hand side of the page to show the operations "O" and inspections "□" undergone by the principal unit of component of the assembly which is the spindle in this case. The time allowed per piece in hours is indicated on the left of each operation. As the inspectors are on time work, there is no specific time allowed for inspections.

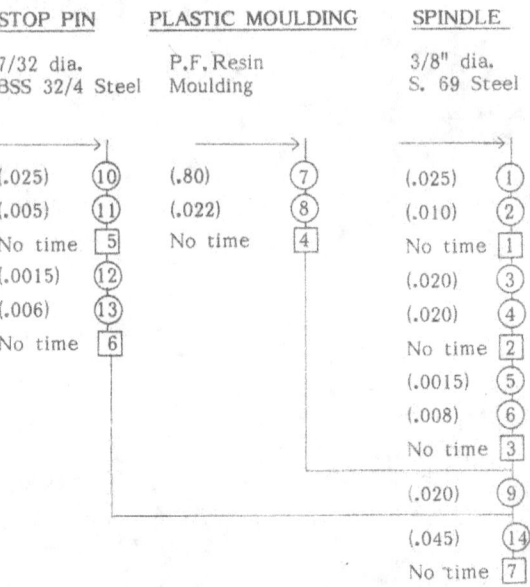

STOP PIN   PLASTIC MOULDING   SPINDLE

7/32 dia.   P.F. Resin   3/8" dia.
BSS 32/4 Steel   Moulding   S. 69 Steel

ASSEMBLING A SWITCH ROTOR

The operations and inspections carried out on the spindle which is made from a 10 mm diameter steel rod are shown below:-

Operation 1
Face, turn, undercut and part off on a capstan lathe (0.025 hr.).

Operation 2
Face opposite end on the same machine (0.010 hr.). After this operation the work is sent to the inspection department for:

Inspection 1
Inspect for dimensions and finish (time unfixed). After inspection the work is sent to the milling section.

Operation 3
Straddle-mill four flats on end on a horizontal miller (0.070 hr.). The work is then sent to the burring bench.

Operation 4
Remove burrs at the burring bench (0.020 hr.). The work is then returned to the inspection department for:

Inspection 2
Final inspection of machining (time unfixed). After this the work goes to the plating shop for:

Operation 5
Degreasing (0.0015 hr.).

Operation 6
Cadmium plating (0.003 hr.). After plating the work goes again to the inspection department for:

Inspection 3
Final check (time unfixed). The plastic moulding with a hole bored concentric with the longitudinal axis is supplied.

Operation 7
Face on both sides, bore the cored hole and ream to size on a capstan lathe (0.080 hr.).

Operation 8
Drill cross-hole (for the stop pin) and burr on two-spindle drill press (0.022 hr.). After this the work goes to the inspection department for:

Inspection 4
Final check on dimension and finish (time unfixed). The work is then passed to the finished-parts store to await withdrawal for assembly.

Operation 9
Assemble the moulding to the small end of the spindle and drill the stop-pin hole right through (0.020 hr.).

After the above activities have been carried out the assembly is ready for the insertion of the stop pin which is made from a 5 mm diameter steel rod; the stop pin is made as follows:-

Operation 10
Turn 2 mm diameter shank, chamber end and part off on capstan lathe (0.025 hr.).

Operation 11
Remove the pin on a linisher (0.005 hr.). The work then goes to the inspection department for:

Inspection 5
Inspect for dimensions and finish (time unfixed). After inspection the work is sent to the plating shop for:

Operation 12
Degreasing (0.0015 hr.).

Operation 13
Cadmium plating (0.005 hr.). After this the work returns to the inspection department for:

Inspection 6
Final check (time unfixed). The work then goes to the finished-parts store where it is withdrawn for:

Operation 14
Stop pin is fitted to assembly and lightly riveted to keep it in position (0.045 hr.).

Inspection 7
The completed assembly undergoes a final inspection (time unfixed). After this it goes to the finished-

parts store.

FLOW PROCESS CHART

The operations process chart is designed to provide an overall picture of a process. But it gives no detail and no record of what takes place between the main operations and inspections. There is no record of movements or lack of movements. Movement in the form of delays in the process or due to storage of material does not add value to the end product but adds to its cost instead. We should know to what extent movement occurs in a process. When this information is added to an outline process chart a flow process chart results.

Description

A flow process chart may be defined as follows:

"A graphic representation of the sequence of all operations, inspections, transportation, delays and storages occurring in the process or procedure, and includes information considered desirable for analysis such as time required and distance moved."

There are two kinds of chart, one based on the activities in a process or procedure related to the people involved in the process and the other based on the activities in a process or procedure related to the material being processed, which are as follows:

[i]   Man-type - depicts the activities of the man or men.
[ii] Material-type - depicts the events as they affect the material being processed.

There should not be confusion over these two kinds of chart. A chart should refer only to the activities of the man or the material and not both at the same time.

Use of Symbols

The two standard symbols utilized in the outline process chart are also utilized to represent the same processes in the flow process chart. However, to indicate movement or lack of movement three additional divisions of activity are included, which are as follows:

Transportation Activity
[i]   The symbol is as follows:

[ii] Transportation is defined as follows;

"A transportation occurs when an object is moved from one place to another, except when such movements are part of the operation or are caused by the operator at the work station during an operation or inspection."

[iii] The predominant result is movement. In a transportation activity the distance moved may be shown besides the symbol. Examples of transportation are moving on a conveyor, flow in a pipe, walking, et al.

Delay Activity
[i]   The symbol is as follows:

D

[ii] A delay is defined thus:

"A delay occurs to an object when conditions except those which intentionally change the physical or chemical characteristics of the object do not permit or require immediate performance of the next planned action."

[iii] The predominant result is interferences or delays. Examples of delay are operator waiting for material, partly processed material which is waiting for further processing, waiting or a lift, et al.

Storage Activity
[i] The symbol is as follows:

▽

[ii] Storage is defined as follows:

"A storage occurs when an object is kept and protected against unauthorized removal."

Examples of storage are finished goods in a warehouse waiting for dispatch, tools awaiting issue, documents filed in a cabinet, et al.

Construction and Conventions used
The construction of a flow process chart is the same as that for an outline process chart, with the same conventions being observed.

Amplification of the Chart
The chart is a means to an end and not an end itself. It should be flexible. Things which can be carried out to increase its value or clarity as a record for subsequent analysis should be encouraged, e.g., adding of colors, hatching, reversing the direction of the transportation symbol, et al.

Shading the "do activities" so that they stand out from other activities in the chart and are given priority during the application of "EXAMINATION" is good practice. Regarding all inspections as "do activities" since they result in decisions affecting quality, further processing and, sometimes, re-processing, is also a good idea. All this has an important effect on the work content of a job.

Conventions used in Process Charting
The following are some of the accepted ways, called "conventions", used for illustrating certain kinds of event. They could be called "charting language".

Charting Language
[i]   When two different activities such as an operation and an inspection occur simultaneously it is alright to blend the symbols together. An example of this might be if an operator moves something with his hand and at the same time check its weight. When an operation and an inspection occur together in this way the numbering convention is retained, with both numbers being placed inside the symbols, as is shown below:

[ii] Introduction of material - a material which is first introduced into a process is indicated by a line and arrow entering from the left, with the nature of the new material written above the line, as is shown below:

Sand

◯ Load

[iii] Change of shape, size or nature during the process - the shape, size or nature of a material may be altered as the result of an operation so that its handling properties from then on are changed, the change being indicated by breaking the chart line at the appropriate place and inserting a brief description of the change, as is illustrated below:

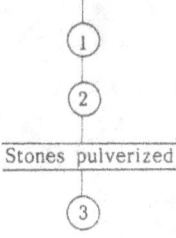

[iv] Entry of subsidiary parts - the final product in many processes is assembled from many components joining the process during its progress. The main processes are charted on the right of the page with the subsidiary processes on the left being joined to the main process by a horizontal line, as is shown below:

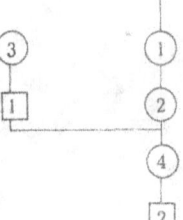

[v]   The charting for rejects for destruction is shown below:

rejects for destruction

[vi]   The charting for rejects for re-processing is shown below:

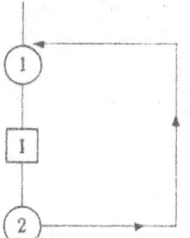

[vii] The charting for repeated activities is shown below:

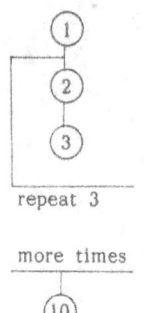

repeat 3

more times

⑩

[viii] Numbering - similar symbols are numbered consecutively from the top downwards. A summary is made at the bottom. A chart which shows what happens to a railway ticket is presented below:

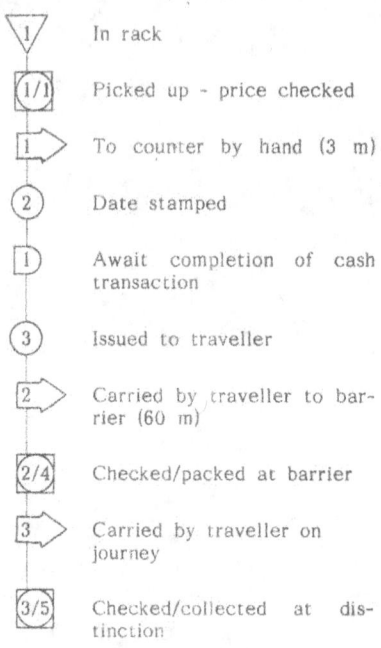

In rack

Picked up - price checked

To counter by hand (3 m)

Date stamped

Await completion of cash transaction

Issued to traveller

Carried by traveller to barrier (60 m)

Checked/packed at barrier

Carried by traveller on journey

Checked/collected at distinction

PRESENT METHOD

⑤

③

▷③ (63 m) rail journey distance
relevant

▽1

1⟩

―――
13

[ix] Notes at the side - it is sometimes a good idea to put notes in brackets beside the activities shown on the chart pinpointing the difficulties. These notes are records of snags and difficulties which should be borne in mind when improving the work method such as:

[a] Difficult or awkward operations.
[b] Distances moved.
[c] Bulky or heavy loads.
[d] Unpleasant conditions, e.g., noise, grease or dirt.
[e] Hazards.

[x] Alternative routes - materials may be divided into portions each receiving different treatment as a result of an operation, in some cases, wherein the main process trunk is divided into the appropriate number of branches with an indication of the proportions following each path, the major parts on the right and the lesser parts on the left, as is illustrated below:

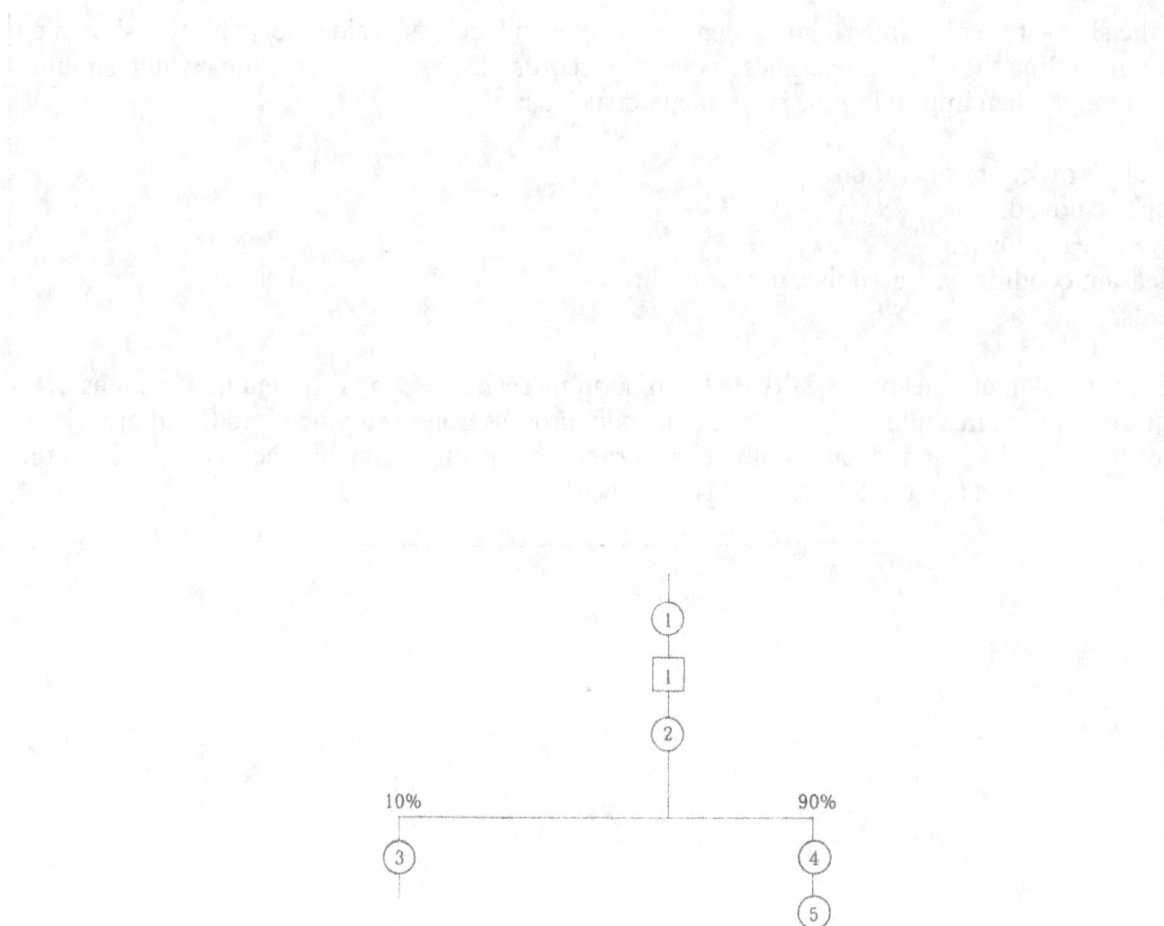

FLOW PROCESS CHART			MAN/MATERIAL/EQUIPMENT		
CHART NO. : SHEET NO. :	ACTIVITY		PRESENT	PROPOSED	SAVING
OPERATION :	OPERATION	○			
PART NAME :	INSPECTION	□			
LOCATION :	TRANSPORT	⇨			
METHOD : PRESENT/PROPOSED	DELAY	D			
CHART BEGINS :	STORAGE	▽			
CHART ENDS :	DISTANCE				
CHARTED BY :	TIME(MAN-HRS)				
DATE :	COST				

DESCRIPTION	QTY	DISTANCE (Km.)	TIME (mins)	SYMBOLS					REMARKS
				○	⇨	D	□	▽	

TWO-HANDED PROCESS CHART

Work confined to a single work-place often involves the use of the hands and arms only. The two-handed process chart is designed to provide a synchronized and graphical representation of the sequence of manual activities of a worker performing such work.

Description

A two-handed process chart may be defined as follows:

"The two-handed process chart is a process chart in which the activities of a worker's hands (or limbs) are recorded in their relationship to one another."

The two-handed process chart is a specialized chart showing the two hands of the operator moving or static in relation to one another.

It is generally used for repetitive operations where one complete cycle of the work will be recorded.

It generally employs the same symbols as other process charts except that the inspection symbols are left out since inspections will be shown as movement of the hands.

For a great variety of assembly, machining and clerical jobs, the two-handed process chart can be applied.

The following symbols are used in the two-handed process chart:

○ Operation is used for the activities of grasp, position, use, release, etc of a tool, component or material.

▷ Transport is used to represent the movement of the hand to or from the work, or a tool, or material.

◗ Delay is used to denote time during which the hand or limb being charted is idle.

▽ Hold (storage) is used to represent the activity of holding work, tool, or material.

Example of A Two-Handed Process Chart
TWO HANDED PROCESS CHART

Chart begins: Hands empty; material in boxes

Chart ends: Completed assembly aside to box.

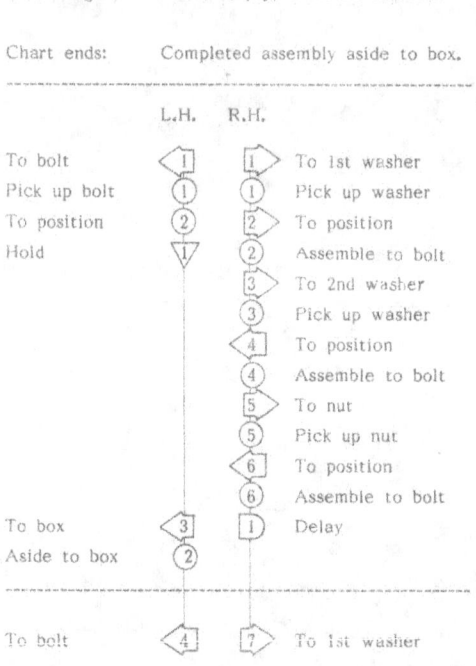

Fig. 1 Two-handed Process Chart Showing Fuller
     Use of Symbols.

TWO- HANDED PROCESS CHART					
PRESENT/PROPOSED		SUMMARY			
OPERATION :		ACTIVITY	PRESENT	PROPOSED	SAVING
CHART BEGINS :		OPERATION ◯			
CHART ENDS :		TRANSPORT ▷			
LOCATION :		DELAY D			
OPERATOR :		STORAGE ▽			
PART NAME :		DISTANT			
CHARTED BY :		TIME(MAN-HRS)			
DATE :		COST			

LEFT-HAND DESCRIPTION	LEFT HAND				RIGHT HAND				RIGHT-HAND DESCRIPTION
	◯	▷	D	▽	◯	▷	D	▽	

MULTIPLE ACTIVITY CHART

The concurrent events affecting one or more operators or an operator and a piece of equipment or material may be recorded on two or more parallel vertical lines.

Description

The following is a definition for the multiple activity chart:

"A multiple activity chart is a chart on which the activities of more than one subject (worker, material or equipment) are each recorded on a common time scale to show their inter-relationship."

The chart shows very clearly the period of ineffective time within the process by using separate vertical columns or bars to represent the activities of different operators or machines against a common time scale, which makes the avoidance of such time by re-arranging the work a very much easier task. The chart should be constructed such that the most important aspect, e.g., cost effectiveness, is given the major emphasis.

Construction of Chart

Normally worker and machine activities are recorded by shading the respective bars. To ensure the chart will be as effective as possible the timing, which can be built up from previous measurements or by direct timing, has to be sufficiently accurate. The activities are then sequentially plotted against the time scale within their own particular bar on the chart. For an example of a multiple activity chart refer to Figure 3.

Uses

The multiple activity chart is especially useful for enabling maintenance and similar work to be organized so that the time expensive equipment is out of commission is reduced to a minimum. (Refer to Figure 2.) It is also a useful method, when organizing team work, of determining the number of machines each worker should take charge. It allows complex processes to be simply recorded for study at leisure.

Example of a Multiple Activity Chart

Figure 2   MULTIPLE ACTIVITY CHART: INSPEC-
            TION OF CATALYST IN A CONVERTER
            (Original Method)

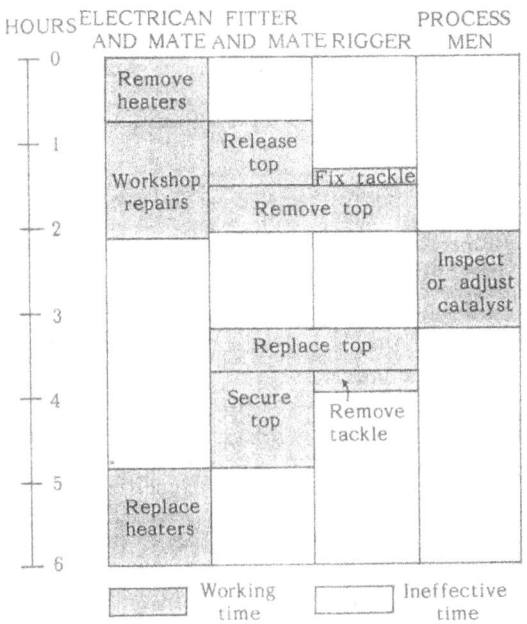

In producing a chart the activities of the different operators or of the different operators and machines are recorded in terms of working time and idle time. These times (which may be in minutes or seconds) may be recorded by using a stopwatch or ordinary wristwatch according to the duration of the various periods of work and idleness. Though extreme accuracy is not necessary, the timing has to be sufficiently accurate for the chart to be effective. The times are then plotted in their respective columns, as is shown in Figure 2 above.

Example of a Multiple Activity Chart

Figure 3   MULTIPLE ACTIVITY CHART: INSPEC-
TION OF CATALYST IN A CONVERTER
(Improved Method)

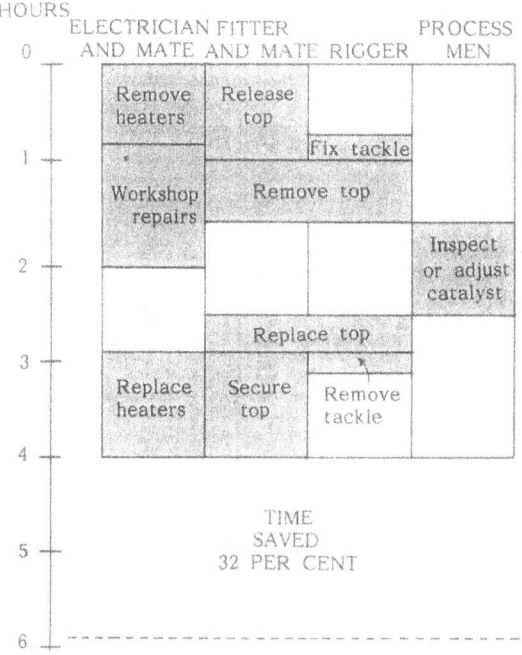

Figure 3 shows that idle time is reduced, the time saved resulting from this simple change being 32 per cent of the total time of the operation.

A simple multiple activity chart can be constructed on any piece of paper which has lines or squares that can be used to form a time scale. Printed or duplicated forms are usually used for this purpose. The general layout of the chart is similar to that of the standard flow process charts, with vertical bars drawn to represent the activities charted.

Questions for Review

1. MANUFACTURE AND ASSEMBLING OF ROCKER ARM - CHARTING PRACTICE

A rocker is made from steel bar of rectangular section ½ inch x ¾ inch. The first operation is punching the drop forging blank on a 40-ton press. The blank is then heated in a muffle furnace to 950°C. It is then formed in one operation on a one-ton drop forge. The spoil is trimmed off in a blanking press. The forging passes to the inspection section where it is electrically tested for flaws and cracks. Then it is routed to the drilling section where it is mounted in a drilling fixture, the boss is drilled, the cheeks are faced using a counter-bore and the rocker bearing is reamed. For this operation, a multi-spindle drilling machine is used. Before the second drilling operation is performed, the machined forgings are inspected. For this the forging is mounted in another drilling fixture which is located in the rocker bearing. The two holes at the ends of the rocker are drilled. One hole is reamed to size while the other is tapped 3/8 inch B.S.F. The operation of "drill two holes, ream one, tap one" is also performed on a multi-spindle machine. After that the forging is passed to the milling section where the two flats are milled on a vertical milling machine.

The completed rocker is sent to the inspection department for a final inspection. It is then routed to the finished parts store where it awaits an assembly order.

The hardened steel tappet is made through case-hardening steel using ½ inch square bright bar stock. The "face, turn and part off" machining operation is performed on a capstan lathe. Using a special fixture, the head is radiused on a grinding machine. After that it is inspected and routed to the heat treatment section for case-hardening. The head is then buffed, after which a final inspection is carried out. Then the part is dispatched to the finished parts store.

The bush is produced from phosphor bronze using ¾ inch diameter bar stock. The "face, turn, drill, ream and part off" machining operation is carried out on a capstan lathe. The parting burr is removed by finishing. After that the part is inspected and routed to the finished parts store.

The adjusting screw is produced from case-hardened steel using ½ inch diameter bright bar stock. The "face, turn thread, form and part off" machining operation is performed on a capstan lathe. After that the ball head is ground by means of a special grinding attachment and a screw-driver slot cut using a slotting saw on a horizontal milling machine. After inspection the part is routed to the heat treatment section for the head to be case-hardened. The head is then buffed, after which a final inspection is carried out before the screw is passed to the finished parts store.

The locknut is produced from 5/16 inch Whitworth hexagonal bright mild steel bar stock in one operation on an automatic screw machine. The part is inspected and routed to the finished parts store.

Assembly is carried out in four operations. The first and second operations consist of running the locknut up to the head of the adjusting screw and screwing the complete sub-assembly into the rocker. In the third operation the phosphor bronze bush is inserted into the rocker forging, this being carried out on a small bench press. Assembling of the hardened steel tappet to the rocker arm, also carried out on a small bench press, is the last operation. Finally, the rocker assembly is inspected and routed to the finished parts store where it awaits a further assembly order.

Draw an outline process chart which shows the complete process. (The answer is provided in Appendix I.)

2. BATTERY RE-CHARGING
Flow Process Chart - Charting Practice

The process of battery charging carried out by Messrs. Tama Motors (P) Ltd is as follows. Customers deposit discharged automobile batteries at the reception bay of the workshop whereupon they obtain a receipt. Three days later, they collect their respective batteries from the delivery bay of the workshop on presenting the receipt and paying the service charges. Each battery is marked with a customer-identification number by the reception clerk who stacks them on a steel shelf.

Mr. Smith collects the chargeable batteries from the reception bay twice a day. He delivers the batteries on a stillage truck to the charging shop, which is 50 meters away from the reception bay.

Each battery is tested cell by cell for voltage and specific gravity at the charging shop. The battery cells are "topped up" to the correct level. The battery is then routed to the charging bench and charged by connecting the terminals to the bus bars. It is tested every four hours for specific gravity while being charged. When charging is completed, which is determined by the electrolyte's specific gravity, the connections to the bus bars are removed. Testing is carried out on the battery for voltage of each individual cell. Then it is placed on a stillage (small wooden platform with low legs).

Draw a flow process chart (material) which shows the complete process. (The answer is provided in Appendix II.)

3. CASE STUDY - ASSEMBLING OF U-BOLT

ASSEMBLY OF U-BOLT CABLE CLAMP
Right and Left Hand Chart

PRESENT METHOD

The U-bolt cable clamp comprises of three different parts: A - the U-bolt, B - the casting, and, C - the hexagonal nuts. The cable clamps are assembled as follows:

With his left hand the operator gets hold of a U-bolt from bin 1 and brings it up in front of him. He then gets hold of a casting from bin 3 with his right hand and attaches it onto the bolt. In a similar manner he grasps two nuts from bin 2 and assembles them onto the threaded ends of the bolt in succession. Then with his right hand he disposes the assembly into bin 4 on his right. The following is the right and left hand chart for this operation:

PRESENT METHOD

	LEFT HAND			RIGHT HAND

LEFT HAND   RIGHT HAND

Reach for U-bolt in bin 1   Carry finished assembly to bin 4
Grasp bolt   Reach for casting in bin 3
Carry bolt to central posi-   Grasp casting
tion   Carry casting to bolt
                                 Position & assembly casting onto
                                 bolt
                                 Reach for 1st nut
                                 Grasp nut from bin 2
Hold bolt   Carry nut to bolt
                                 Position & assemble nut onto
                                 bolt
                                 Reach for 2nd nut
                                 Grasp nut from bin 2
                                 Carry nut to bolt
                                 Position & assemble nut onto bolt
Released finished assem-   Grasp finished assembly
bly to right hand

Develop a better method for the assembling of the cable clamps. Make a right and left hand chart of the proposed method. (The answer is provided in Appendix III.)

7 EXAMINATION AND DEVELOPMENT

INTRODUCTION

Further timings should be obtained by one of the techniques of work measurement when recorded facts are found to be insufficiently accurate.

It is important to keep to a set plan when studying the recorded facts of a process. When examining the recorded facts attention should be focused on individual aspects of an activity. The next aspect should be investigated only after full consultation with the people concerned has established the true facts and reasons underlying one aspect of the process. With this full knowledge of the underlying reasons, the work study officer can review the process as a whole, and, using the same system of full consultation with those who may be able to provide useful knowledge and ideas, look for the available alternatives.

APPROACH

The results which can be achieved through this examination will depend on the attitude of the work study officer and his ability to obtain relevant information from the many available sources. However, the following points should be borne in mind:-

[a] Facts should be examined as they are, not as they appear to be, should be or are said to be.
[b] Preconceived ideas, which often color the interpretation of facts, should be avoided.
[c] All aspects of the problem should be approached with a challenging and skeptical attitude, with every detail logically examined and no answer accepted unless it has been proved correct.
[d] Hasty judgments should be avoided.
[e] Details should be closely and thoroughly examined.
[f] Hunches, which should be immediately recorded when they appear, should be delegated to the appropriate place in the investigation.
[g] Unless all the undesirable features of the existing work method have been exposed by systematic examination, new work methods should not be considered.

The stated objectives of the work study should be borne in mind all the time. Improvements in work method will result in the elimination of waste of both materials and time.

CLASSIFICATION OF ACTIVITIES

It may be possible to reduce work by eliminating certain operations or changing their sequence, though transports and delays may on the surface appear to provide the greatest scope for improvement. This will affect transports and delays automatically. An examination of the operations in the process should initially be carried out according to their order of importance in the overall process. If the operations have been divided into "make ready", "do" and "put away" classifications on the indicated lines, the elimination of any of the "do" operations will result in any "make ready" and "put away" operations which are connected with them, as well as the corresponding transports and delays, to be automatically eliminated.

To prepare the material or work-place and set in position the material ready to be worked on, "make ready" activities are needed.

"Do" operations are operations that represent the actual performance of work on material and that result in change in the characteristics or properties of the material. Under certain circumstances, e.g., in maintenance work, the change may be in the condition of equipment or plant. All inspections are regarded as "do" activities as they usually result in important decisions which affect further processing, reprocessing or rejection from a process. "Put away" activities are operations which are concerned with placing aside or cleaning-up following another operation or inspection.

QUESTIONING TECHNIQUE

The questioning technique is the way by which the critical examination is carried out; each activity is subjected in turn to a systematic and progressive series of questioning.

In the application of the questioning technique there is a likelihood that many will dislike suggestions that the way they work is incorrect and resist change. The work study officer should be diplomatic in his approach and obtain the confidence and cooperation of the people concerned. In other words, he should understand the human factors involved.

When questioning the existing work method, the following aspects should be looked into:-

[i]   The work done.
[ii] Where it is done.
[iii] When it is done.
[iv] By whom it is done.
[v]   How it is done.

For each of the questions, find out why and what achievement, person, place, time or way will be better, as is illustrated below:-

THE PRIMARY QUESTIONS

The sequence of questioning used follows a well-established pattern involving the following:-

the PURPOSE for which the activities are undertaken
the PLACE at which the activities are undertaken
the SEQUENCE in which the activities are undertaken
the PERSON by whom the activities are undertaken
the MEANS by which the activities are undertaken
with a view to ELIMINATING, COMBINING, REARRANGING or SIMPLIFYING those   activities

The PURPOSE, PLACE, SEQUENCE, PERSON and MEANS of every recorded activity are each systematically queried. A reason for each reply is sought.

The primary questions are as follows:-

PURPOSE: WHAT is actually done? WHY is the activity necessary at all?
PLACE: WHERE is it being done? WHY is it done at that particular place?
SEQUENCE: WHEN is it done? WHY is it done at that particular time?
PERSON: WHO is doing it? WHY is it done by that particular person?
MEANS: HOW is it being done? WHY is it being done in that particular way?

[ELIMINATE unnecessary parts of the job. COMBINE wherever possible or REARRANGE the sequence of operations for more effective results. SIMPLIFY the operation.]

It is important not to confuse the questions and answers pertaining to "purpose" and "means" since the objective of the primary questions is to ensure that every aspect of an existing method is clearly understood.

For example, when considering the operation of tying a parcel with string, the question "What is achieved?" under PURPOSE is given the answer "The parcel is fastened." and not "The parcel is tied up with string." Similarly, in the questioning for the operation of planning a certain length of timber to ½ inch thickness, the question "What is achieved?" under PURPOSE is given the answer "A length of timber, 36 inches x 6 inches, is reduced to ½ inch thickness." The question "How is it done?" might be given the answer "By planning with a rotating cutter block on a 9 inch Wadkin surface planing machine."

The primary questions provide the background of events and establish whether the existing procedures are based on firm reasoning, with any part of the work which is unnecessary or inefficient with respect to PLACE, SEQUENCE, PERSON or MEANS.

THE SECONDARY QUESTIONS

In the second stage of the questioning technique, the secondary questions come into play. The answers to the primary questions are now subjected to further questioning to determine whether the possible alternatives of PLACE, SEQUENCE, PERSONS and/or MEANS are practicable and preferable as a means of improvement over the existing work method.

Having already asked about every activity recorded, what is done and why it is done, the work study practitioner goes on to enquire what else might be done, and, what should be done. The answers already obtained for PLACE, SEQUENC, PERSON and MEANS are similarly subjected to further enquiry.

The secondary questions should cover the following:-

PURPOSE: What is done? Why is it done? What else might be done? What should be done?
PLACE: Where is it done? Why is it done there? Where else might it be done? Where should it be done?
SEQUENCE: When is it done? Why is it done then? When might it be done? When should it be done?
PERSON: Who does it? Why does that person do it? Who else might do it? Who should do it?
MEANS: How is it done? Why is it done that way? How else might it be done? How should it be done?

The answers to the secondary questions will be the pointers for improvement. The following considerations are important:-

[i] When there is doubt about the purpose of the activity it is most important to consider whether it can be eliminated entirely.
[ii] If the activity is essential, how it is to be done is next looked into as the means will frequently determine where it is to be done and who does it. For example, if two pieces of metal were to be fastened the how determines the who. If welding is considered best, a welder has to do the work. Which welder to do the work can be determined once the means is established.
[iii] It is important to consider whether the activity can be modified or combined with other activities as improvements are often obtained by combining work, or, changing the place where work is carried out, the sequence in which activities are performed or the persons carrying out these activities.
[iv] Examination Procedure
The examination procedure involves two sets of detailed questions, namely, the primary questions to elicit the facts and the reasons behind them and the secondary questions to determine the alternatives and ultimately the means of improvement. The questions are classified into several headings - the purpose of the operation, the place where the operation is carried out and the means by which the operation is carried out. Both the primary and secondary questions are asked for each aspect of the work before going on to the next. If the detailed questioning through both the primary and secondary

questions does not establish a purpose for an operation, it is unnecessary to inquire about any other aspect of the operation.

After the full examination of the operation, other activities which still appear to be necessary can be subjected to the same examination procedure.

THE QUESTIONING TECHNIQUE

(Note all ideas)

The Present Facts		C-Alternatives
A-Key Operation	B - Why	
Purpose - What is achieved?	Is it necessary? Yes/No If yes - why?	What else could be done?
Place - where is it done?	Why there	Where else could it be done?
Sequence - when is it done?	Why then?	When else could it be done?
Person - who does it?	Why that person?	Who else would do it?
Means : How is it done?	Why that way?	How else would it be done?

Fig. 6 - Show Critical Examination Sheet Headings.

EXAMPLE OF CRITICAL EXAMINATION

The following is an example of what happens when someone mows the lawn with a power mower:

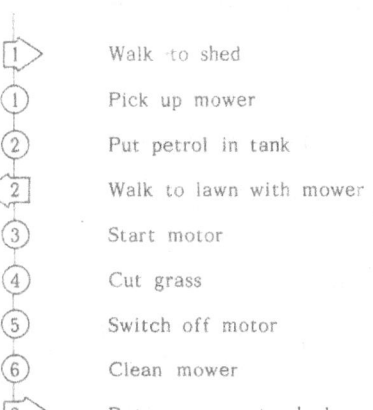

▷1	Walk to shed
①1	Pick up mower
②2	Put petrol in tank
◁2	Walk to lawn with mower
③3	Start motor
④4	Cut grass
⑤5	Switch off motor
⑥6	Clean mower
▷3	Return mower to shed

"Cut the grass" is the "Do" activity here. If the need for this activity is eliminated all the other related "make ready" activities, delays and transportations will also be eliminated. This is the reason the "Do" activities should be challenged first. A "Do" activity is not any operation, e.g., "fetch mower" and "return mower to shed" are operations but they do not bring about changes in the characteristics or properties of anything and are thus not "Do" operations.

"Cut the grass" is the first "Do" operation.

The first question to ask about it is "What is achieved?". The answer is "The grass is cut.".

Is it necessary? "Yes."

Why? "Because the grass has grown too long."

What other achievement would be better? "Prevent growth." or "Eliminate grass." would be possible alternatives.

Where is it done? "On the lawn."

Why there? "Because that is where the grass is."

What other place would be better? "There is no possible alternative."

When is it done? "When the grass becomes unsightly."

Why then? "So that its appearance could be improved."

What other time would be better? "More often, because when the grass is long and unsightly it clogs the mower."

Who cuts the grass? "Father."

Why that person? "He has always cut the grass."

What other person would be better? "Possibly a son or daughter, or, neighbors on a rotating basis with father taking his turn."

How is the work carried out? "By using a mower with a motor that runs on petrol and emptying the grass box every 20 meters."

Why in that way? "Because we have such a mower."

What other better way of carrying out the work is there? "By using an electric powered rotary mower, or, cutting the grass more often such that it would be possible to let the grass lie on the lawn."

The following are all the ideas that have been generated so far:-

[i]   The lawn could be eliminated.
[ii] A slower growing variety of grass could be introduced.
[iii] A son or daughter could carry out the work of cutting the grass.
[iv] If the grass could be cut more often:

   [a] It would be an easier task.
   [b] The grass would not clog the mower.
   [c] Fuelling would be unnecessary.
   [d] Storage of fuel, which is a potential fire hazard, would be unnecessary.
   [e] Collecting the cuttings with all the problems of transporting them and disposing them would be
       unnecessary, et al.

v] Neighbors could take turns to cut the grass. As much effort is required in getting the mower out and in cleaning it after cutting the grass, the frequency of doing all this would be lessened if the whole lawn could be mowed in one go instead of by piece-meal.

Go on to challenge other operations such as "make ready", "put away" and "inspection" if there is still the need for them, after all the "Do" operations have been challenged (as is described above). Proceed next to the transportations, delays and storages and decide whether they need to be challenged in the light of the challenging that has been carried out and the ideas that have already been produced.

At this stage, note down all the ideas. No idea should be dismissed simply because it does not fit in with an evolving plan. Hold on to these ideas as they may be able to help bring improvement.

When thinking about better ideas, the following questions and considerations are important:-

[a] Elimination of part of or the whole of the operation.
[b] Re-arrangement of the order in which operations are carried out.

[c] Combination of one operation with another operation.
[d] Simplification of the question - utilize an alternative machine.
[e] Re-designing of the item so that part of the operation could be eliminated.
[f] Substitution by a standard purchased part.

[1] Questioning of the Materials Used and the Materials Handling Methods
Consider the following:

[a] Substitution with less expensive material.
[b] Utilization of scrap.
[c] Revision of specification to allow the use of more readily available material.
[d] Consideration of quality and type of material.
[e] Improving the shape or condition of material so that handling is easier.
[f] Pre-positioning of material at the work area and disposing of the finished item.
[g] Where possible gravity feed to the handler of the material should be arranged and the delivery point should be close to the point of use and arranged such that further lifting or repositioning is unnecessary.
[h] Disposal should be arranged as a drop movement wherein the material is released in the position in which it has been worked on and at the point where work ceases.

[2] Questioning of Machines, Tools and Equipment Employed
The main question to be considered now is whether the physical equipment and arrangement are the most suitable for the item that is being manufactured. The following could be considered:

[a] The most suitable machine should be utilized for the job.
[b] The idle time of the machine should be utilized productively.
[c] Possible improvement for the set-up of machines should be investigated.
[d] Manual operations should be reduced to the minimum by the provision of jigs and stops.
[e] Operating levers should be arranged according to the sequence of use.

[3] Questioning of the Worksite Layout
The conditions wherein the greatest economy in motion may be achieved are considered here, e.g., the question of whether the worksite layout is the most effective. The following could be considered:

[a] Supply all essential items in the worksite area.
[b] Effective utilization of both hands.
[c] Elimination of holding by the hands through the introduction of jigs and fixtures.

[d] Utilize gravity feed hoppers.

[e] Minimize number of handlings and distances traveled.

[f] Reduce hand movements by providing foot-operated devices.

[g] Eliminate duplication of motions and back-tracking.

[h] Height:

    [i] The height of any chair or stool provided and of working surface should be such that a a worker could either sit or stand.

    [ii] It should be possible to adjust the height.

    [iii] Whenever possible, bending and lifting should be avoided.

[4] Questioning of Operator Training and Working Methods

Investigate whether the following may be carried out:

[a] Elimination of all motions that are unnecessary.

[b] Wherever possible, combination of motions.

[c] Re-arrangement of sequences of motions.

[d] Simplification of motions.

[e] Standardization of required motions.

[f] Utilization of motions that are most practicable.

[g] Establishment of motions which are safe.

[h] Choose appropriate equipment and tools:

    [i] Each piece of equipment or tool should be placed in a fixed location when not in use.

    [ii] The equipment or tool should be placed in front of and within the grasp of the worker.

    [iii] Tools should be placed at the same level and in the order of use.

    [iv] Tool handles, et al., should be located such that they could be manipulated with the smallest change in body motions and with the greatest mechanical advantage.

    [v] Wherever possible, two or more tools should be combined into one.

    [vi] Where economically justified, tools should be power-operated.

    [vii] To reduce the load on the worker, heavy tools should be provided with a counter-balance.

    [viii] Handles should be designed in such a way that as much of the surface of the hand is in contact with the handle as possible.

[i] Where possible, jigs and fixtures should be utilized to assist in the operation.

[j] Where they would be helpful, dials and gauges should be used:

    [i] Control dials and gauges should be located together on one panel in front of the worker and in positions according to the order in which they have to be read.

    [ii] The control limits or the points at which a worker has to take action should be indicated on the

instrument card.

[iii] The place where the written record is produced should be adjacent to the source of information.

[iv] Sufficient illumination should be provided and should be directed at the instrument.

[v] The method of illumination should not cause eyestrain to the worker.

[k] Whenever possible, equipment controls should be foot-operated but if hand operation is necessary then operation by the left hand is preferable to the right hand when the right hand could be better employed.

All these questions and ideas are not exhaustive but are merely aids in investigating the possibilities of improvement in each of the main categories. They basically boil down to the consideration of questions prompted by the six questions:

What?

Where?

When?

Who?

Why?

How?

DEVELOPMENT

In order to develop the new work method, the diagram shown below should be extended to include a "should" column D on the right:

The Questioning Technique

(Note all ideas)

A. Key Questions	B. Why?	C. Better Ideas	D. Should
			What should be achieved?
			Where should it be achieved
			When should it be achieved?
			Who should achieve it?
			How should it be achieved?

Ideas should be reviewed and trends should be noted. In reviewing ideas one has to look at all the suggestions and notes one has made before and during the questioning process. Noting trends involves identifying the trends or lines along which one's mind and the minds of others have been working.

For instance, in the above-mentioned lawn mowing job the trends had been towards changing the lawn mower and delegating the job. In considering better methods there may be two or three trends or main thought patterns.

Next, the following should be looked into:-

Eliminate
Simplify
Combine
Re-arrange

The above concepts apply not only to the existing work method but also to the embryonic new work methods that have been forming in one's mind.

The elimination of the work of collecting and disposing of cut grass might be suggested by "eliminate". The possible purchase of an electric lawn mower of the same type as the neighbors' wherein parts are interchangeable might be suggested by "simplify". Working with neighbors on the problems by taking turns to do the lawn-mowing or some other cooperative measures might be suggested by "combine". Re-arranging the sequence of a job or the location of equipment, e.g., the place where the lawn mower is kept for convenience, might be suggested by "re-arrange".

After all this, the safety of the new work method should be looked into. The use of an electric lawn mower comes with possible hazards such as electric shocks, e.g., someone may be electrocuted if he does not know that the power is connected. Safety ideas could be incorporated, e.g., having printed instructions near switches and properly instructing those persons who handle the lawn mower for the first time.

Finally, chart the new method, which is not always necessary but can sometimes help when considering the next item in Development.

SUBMIT FOR APPROVAL
When one submits the new work method for approval one should consider how one is going to sell it, e.g., what are the benefits and advantages to the person who is going to approve it, what kind of person he is, what are the things that motivate him, et al.

He may be concerned with the following:-

Safety
Efficiency
Cost saving
Prospect of greater cooperation and teamwork
Human relations
Progressive image
Competitiveness

To win him over, one should use the levers which one knows he will respond to. One should try to be specific about anything which can be measured, such as the time taken or costs, instead of vaguely claiming it as improvement.

If one presents ideas on paper they should be concise, legible and clear. One should avoid presenting a long document as it will put many people off; one can always provide more detail if asked.

Questions for Review
1. Propose a job that you are familiar with and implement the procedure of Critical Examination.

8 INSTALLATION AND MAINTENANCE

INTRODUCTION TO INSTALLATION

Implementing the basic procedure of work study will eventually lead to the stage whereby the agreed work method is ready to be put into practice. Installation of the work method is not simple and will need the active support of everyone concerned. For installing the work method, an opportune time should be chosen. Until sufficient preparation has been made, no attempt should be made to introduce the changes. For any worker who may be displaced as a result of the new work method, alternative work should be found for him.

When it is certain that the people concerned have confidence in it and will support it, the new work method can be installed.

The installation should be carried in two stages as follows:-

[1] Preparation
[2] Installation

Under both the headings above, there are many considerations, the extent to which they will apply in particular cases being dependent on the nature and scope of the changes to be made.

The work study department will always be actively responsible for some of the work involved, while for the remainder of the work in which executive action has to be taken by one or more of the service or operating departments concerned the work study officer's role will be one of advisor and coordinator. The work of installation must always be a cooperative effort. It is the work study department's function to ensure that the scheme is implemented according to the recommendations made and accepted in the report. For success, the confidence of management in the possibilities of work study and in the competence of its practitioners is essential.

PREPARATION

Before the actual installation takes place, detailed preparation should be made. This preparation can be broken down into three stages as follows:-

[1] Plan
[2] Arrange
[3] Rehearse

Plan
A general program for the installation should be produced, for example:

[a] One person only should be responsible during the installation of the work method and everyone should know who that person is. He may have to delegate his authority but everyone should know that he holds the final responsibility and authority.
[b] Actual dates should be fixed for each stage if the installation is to take place in stages which have not already been announced before the preparation of the report. These dates should be selected such that they are convenient both for the people responsible for or affected by the installations and for the process itself.
[c] Copies of any time-table for the installation should be in line with the dates selected for each stage. It may be advisable for the time-table to be in much greater detail than the form in which it appeared in the report.

Arrange
The necessary detailed arrangements should be made, for example:

[a] Make detailed check of all layouts to ensure that all the necessary plant, equipment and tools are available and that services are laid on. Ensure that everything which has been ordered will be ready when required.
[b] Arrange for the running-down of old stocks and the building-up of necessary new stocks of materials ahead of the installation.
[c] Check whether all supplies and services are available at all times.
[d] For purposes of control and comparison, set up any additional or new clerical records which may be required.
[e] If there are changes in hours of work, e.g., from day to shift work, ensure that auxiliary services such as transport, canteens, et al., are informed.
[f] Select carefully the number of workers with the required ability for the new work method. This is especially important when there is teamwork involved. Try to prevent difficulties which may arise between the workers selected for the new work method and the workers who are still with the old work method.
[g] Provide the necessary training, which should not be skimped and should be thorough, especially where teamwork is needed. Training should take place away from the production floor, preferably in an independent training department, where possible.
[h] Anticipate wage and payment problems and settle them well ahead of time. Ensure that the wages and costing departments know when the new work method is to be installed.
[i] Update everyone concerned on the plans and time-table for the installation.

Rehearse

It is often advantageous to give the improved work method a trial run. The following aspects should be looked into:

[a] The rehearsal usually takes place while the old work method is still in operation and probably has to be conducted outside normal working hours so that normal production will not be affected.
[b] In order that proper quality standards are maintained, ensure that all inspections have been allowed for.
[c] All departments that are affected by the change should be represented at the rehearsal.

INSTALLATION

Actual installation can be carried out after all detailed preparations have been made and a successful rehearsal has been held. The following should be looked into:-

[1] The physical aspect of the switch over of work methods can generally be implemented outside normal working hours. A week-end or holiday is often a suitable opportunity as there will be sufficient time to alter layouts and install plant and equipment without affecting normal production.
[2] Since the first few days of operating the new work method are critical, during this period very close supervision should be carried out. This extra supervision should continue till all workers are much familiar with their job. Meetings should be held each day with the supervisors concerned so that progress can be discussed.
[3] Despite the effort to ensure that the work method is the best and most practical, there may be some aspect of it which does not turn out as well in practice as had been anticipated whereby some modification may be necessary. Whenever changes are made to the work method, the operating instructions should be modified accordingly.
[4] Just after the installation a close look-out should be kept for any prejudice against aspects of the work method among the workers and the reasons for this prejudice should be sought immediately.
[5] Any weak links should be bolstered by extra training. Unsuitable personnel may be replaced and more labor may be temporarily allotted to parts of the work method.
[6] The effect which the new work method is having on stocks should be closely monitored.
[7] Tact and restraint should be exercised throughout the period of the installation and ample credit should be given where due.

When the department heads concerned are satisfied that the new work method is running smoothly and are prepared to accept it as a going concern, the installation can be regarded as complete.

MAINTENANCE
<u>Maintaining the New Method</u>
After an improved work method has been installed and is operating well, measures should be taken, through supervision, to ensure that the work method adheres to the authorized procedure. Changes which are unofficially made often result in inefficiency, such as deterioration in safety and quality standards and duplication of work. It is especially important in the period immediately following the installation to ensure that defects of the old work method are not revived by habit in the new work method.

The long-term reactions to the improvements of both the workers using the work method and the other sections affected by the changes when they were made should be noted. Much can be gained from their reactions to the improvements - these reactions may provide clues that will disclose still further possibilities for improvement.

There is an excellent opportunity for making the organization "method conscious" when work methods are reviewed. Work study can be applied on a much wider scale within the organization with the active support and cooperation of all concerned when the benefits to be obtained from it are fully appreciated.

The conditions are bound to change from time to time which may mean that some of the assumptions on which the improved work method is based are no more valid. The labor for manning the improved work method should be allotted on the basis of the work content of the work method as is set out in the operating instructions. Thus any changes may alter the balance between the work content and the labor allotted for implementing the work method. To make allowances for any changes, the work method should be reviewed at intervals.

Changes may be the result of any of the following main causes:-
[1] For good reason there can be deliberate and discernable changes and improvements to the work
    method.
[2] Minor innovations may be introduced by the supervisors or the workers.
[3] Changes may be the result of suggestion schemes.

The work study officer should be given advanced notice so that he can make the necessary allowances for the effect which changes will have on the operation of the work method when changes in work method are deliberately introduced by management. There should not be difficulties if there is communication between the work study section and the other sections concerned. In the sections concerned a copy of the operating instructions should always be available.

Minor changes are liable to creep into the work method from time to time, resulting in a gradual drift away from the authorized work method. The most effective way of exposing these changes of work method is to have a regular review of the current practice.

Reviewing the Method

How frequently a work method should be reviewed will depend on the nature of the work. The main purpose of the review is to find out whether there are any discrepancies between the authorized work method as specified in the operating instructions and the current practice at the time of the review.

Any variations in the work method should be investigated. Changes which have occurred for valid reasons should be accepted, with the operating instructions amended accordingly. Operating instructions should be amended the soonest possible once changes in the work method are authorized. The work study officer should ensure that credit for the improvement in the work method is given where it is due.

However, if the review shows that there are undesirable variations in the work method the work study officer should rectify the situation.

Questions for Review

1. Select a job you are familiar with, apply work study techniques and attempt to find a better method of doing it.

Describe how you will implement and install the improved work method.

9 THE ROLE OF THE SUPERVISOR IN THE STUDY PROCESS

INTRODUCTION

The supervisor is frequently greatly affected by the application of work study. The introduction of the improved work techniques by the supervisor calls for the ability in utilizing and organizing staff on his part, an activity which he has probably confidently performed for many years. Cooperation is important to the successful application of the improved work techniques and in establishing a good working relationship between the supervisor and the workers. Work study is a continual process. The supervisor is mainly responsible for its lasting effectiveness after the initial application.

ROLE OF THE SUPERVISOR

The role of the supervisor is often altered when work method improvement is introduced. It is of utmost importance that the work study officer helps to develop and equip the supervisor so that the latter can meet the changed demands that have been placed on him.

As the supervisor is at the front line dealing with the workers, he should be alerted to the impending change of the work method concerned. He should also cooperate with the work study practitioner.

On the other hand, the work study practitioner in carrying his duties should seek the benefit of the skill and experience of the supervisor.

The supervisor generally holds his position by virtue of these factors. Apart from the other direct benefits, work method improvement cannot be fully effective without them.

It is highly important that the supervisor is kept informed of the progress of the work of the work study practitioner in those areas for which the supervisor is responsible.

WORK METHOD IMPROVEMENT AND WORKERS

The new work techniques cannot be fully effective if the cooperation of those whose work is in the area affected by the work method improvement is absent. The attitude of workers to work method improvement or to any other management technique will depend on their relations with the management of the organization. The work study practitioner in implementing the improved work method should be able to establish a working relationship with the workers based on mutual respect.

As the cooperation of the workers involved is necessary for the success of the work method improvement it is highly important that their supervisors encourage the cooperation. In particular the supervisor should:

[i]   Always provide the true purpose of the project and not mislead.
[ii] Explain as simply as possible the objectives of the project and the means of pursuing them.
[iii] Explain the likely, foreseeable effect of the project on employment conditions.

Questions for Review
1. Who are the key persons who are involved in work method improvement?
2. How does the supervisor convince the workers of the benefits that can be gained from work method improvement?
3. In the process of work method improvement why is the role of the supervisor important?

10 TIPS ON HOW WORK STUDY CAN BE EFFICIENTLY CARRIED OUT

INTRODUCTION
Besides knowledge of work study techniques, the application of sound common sense is also important for carrying out a successful work study program. Work study should be properly carried out.

REACTIONS TO APPLICATION OF WORK STUDY
Work study is now a specialist function in industry and in other fields and is regarded as an essential management function. Though the main objective of work study is to improve the existing method of doing things by bringing about change, attention should be paid to the reactions of all levels of employees who experience these effects.

Work study should be included in the normal process of management and should not be left just to the specialists. It should not be started and carried out in a haphazard manner and should be continuous. In order that managers have proper executive control over their application and a full appreciation of their potential benefits, they should have sufficient knowledge of work study techniques.

To a greater or less extent every employee is interested in his own personal gain. Unless work study techniques are properly applied, considerable resistance at all levels within the organization can be expected. The trade union has first to be contended with. At the outset of the application of work study many companies go to great trouble to bring the trade unions into the picture at the earliest possible time. There should be no hesitation in discussing honestly with the unions all the problems involved and the possible effects and repercussions. For the successful application of work study techniques, cooperation from the unions is of great importance. Most unions will appreciate the immediate benefits of work study such as eliminating drudgery, frustration and unhealthy working conditions, providing the opportunity for higher earnings and increasing the profitability of the company concerned and thus boosting the nation's economy as a whole. There will naturally be some fear that some employees will be retrenched as a result of work or organizational re-organization. This is the reason why it is all the more important that the fear of the employees is alleviated through their union. Other means such as company newsletters and explanatory booklets will of course also play an important part.

Managers may also feel threatened by the findings of the work study practitioner, particularly when well-established routines are going to be upset. Thus, the support of top management is necessary, without which any effort to initiate work study is bound to end in failure. As the managers may feel incompetent or threatened by the evident improvements resulting from the work study, some senior person should explain that it is a principle of the organization that there will be no recriminations or fault-finding as a result of the facts which have emerged from the work study. Managers should be made aware that work study is a management tool which ought to be accepted by all and not be viewed with suspicion. It is of the utmost importance that lower level management personnel such as charge-hands, foremen and supervisors are as closely involved with work study as possible as they are closest to the workers and can act as an excellent link between the work study officer and the workers. They convey to workers the details of what management requests to be done and are responsible to a great extent for implementing the program of work on a daily, weekly or monthly basis, as well as for output and quality of production, the proper utilization of labor and raw materials, maintaining safety standards, the methods used for carrying out the work, the training of new workers and the retraining of existing workers.

Everyone within the organization where work study is undertaken should be convinced that:

[i]   It is necessary to reduce manufacturing costs.
[ii] There are more advantages gained from systematic method study than from haphazard attempts at work method improvement.
[iii] Advantages can be gained by measuring work rather than relying on labor requirements sanctioned by custom and established possibly a long time ago.
[iv] All workers will be paid fairly for additional work.
[v]   Incentive and bonus schemes will be implemented fairly.

The work study practitioner should carry out his job in a cautious, tactful manner. Every established organization has its own culture. Work study by its very nature is a challenge to the customs and practices of the organization, seeking to change some aspects of the organization based on facts which when proven may be much different from what people believed was happening, e.g., the discovery of inefficiencies and wastage of time, effort and material which were not at all evident before the work study was carried out. It is important that pointers to possible improvement should not be used as a means of blaming those who have been doing their best with insufficient information; these people should not be blamed for not having thought of the desirable changes and should be provided patient and clear exposition of the reasons why the changes are desirable.

If work method improvements were to be really effective concern for those involved is necessary.

HOW WORK STUDY SHOULD BE INTRODUCED

When introducing work study, two steps can be taken. First of all, the work study officer should make everyone aware of the management's intentions and create a favorable climate by educating management personnel and selected staff and their representatives. Secondly, work study should be introduced to various sections of the organization, which is a lengthy process taking possibly up to a year before the first application happens.

The support of the highest authority in the organization, e.g., the board of directors in an industrial company, is necessary if work study were to be successfully carried out. If such support were not available the introduction of work study should be deferred till it is. With the support from the board of directors confirmed, the next person to be directly concerned is likely to be the senior official in one locality, e.g., the works manager of the factory or the office manager of an insurance company where work study is to be introduced. This senior official should be strongly committed to and determined on action instead of just giving the project lip service which is unfortunately far too common. He should have a great appreciation of work study and should attend a good appreciation course on it. He should meet up with others of like status and thereby realize the basic similarity of their mutual problems. No amount of reading and informal discussion can replace this sharing of experiences and ideas whereby a better understanding of work study can be achieved.

There are courses run by various organizations serving industry. Such courses may be conducted internally if the organization in question is sufficiently large. These courses should last for one week though two weeks would be just fine and should be of high quality.

If the education of the manager is inadequate trouble, difficulties and a distorted picture of what work study can achieve will be likely to occur, e.g., the case of the middle managers who can directly or indirectly influence the application of work study. The application of sound common sense, which is indeed a very difficult thing to teach, is highly important.

MIDDLE MANAGERS AND SUPERVISORS

The works manager or office manager should keep his middle managers and supervisors, as well as other related personnel, informed of the organization's intention to introduce work study, providing them a broad overview and letting them know that they will be receiving further information and training in due course. A great deal depends on the manager's handling of this phase - he has to build up the confidence of his management team and instill in them a spirit of enthusiasm for work study. He should strongly emphasize that if there were changes and improvements none of these will be used to criticize the present or past activities of managers, which is vital to the morale and confidence of the supervisors and middle managers.

UNION OFFICIALS AND MEMBERS

The management's intention to introduce work study in the organization should be discussed with the trade unions at the earliest possible moment. There is everything to gain by this action and much to lose by delaying this consultation. The manager or an appropriate company official should hold an informal meeting with the local union officials, with some of his staff who are immediately concerned in attendance. It should be clear that there will be no commitment asked for or given, but that the meeting is a firm and practical demonstration that the management has the intention to consult with the unions at all stages as more facts are established and progress is made. It should be made known that dates are not yet conformed, staff will have to be appointed and trained and it may be some time before demonstrable progress is made.

Prior to convening the meeting the management should first have a clear picture of the labor problems involved and have prepared the framework of a policy that will guide their treatment of individual cases. As all eventualities cannot be foreseen, a well-conceived policy should be flexible and far-sighted enough to avoid embarrassing precedents and disputes due to inconsistent interpretation. Redundancy is an important point of concern; at the outset policy on this should be very clear. Trade unions should be kept informed about all this and other aspects at practically the same time as the middle managers and supervisors to prevent unofficial information circulating within the organization and distortion of information.

All union shop stewards and/or joint consultation representatives likely to be involved in the work study project should be provided clear and practical help in order to gain an understanding of work study. All this can be carried out by the manager at a meeting, or, meetings. Facilities can also be provided for the local trade union officials to brief their shop stewards. Thereafter notices should be posted throughout the organization to inform all of the management's intention to introduce work study. Middle managers and supervisors should be encouraged to discuss work study with their staff.

TIME FOR MAKING ANNOUNCEMENTS
Great effort should be made in building trust and confidence throughout the organization, which can only be possible when everyone is convinced that work study will benefit all. There is however the possibility of the work study project being sabotaged by the ill-intentioned, which should be obviated. To prevent sabotage, the manager concerned should provide carefully timed and demonstrably honest explanation of work study to the staff.

Middle managers, supervisors, local trade union officials, shop stewards and workers should all be officially informed about the introduction of work study on the same day and preferably in the same working period, e.g., the afternoon. For instance, just after lunch, in the early afternoon, middle managers and supervisors can be officially informed at a meeting, followed by the local trade union officials at mid-afternoon and the shop stewards and/or joint consultation representatives in the late afternoon, with notices put up throughout the organization just before dismissal time. This example may represent a tight and difficult schedule but keeping everyone in the organization officially informed about the introduction of work study is much worth the trouble.

After this, efforts can be made to hire or acquire the work study personnel considered necessary for the implementation of the work study program. Unless the management's intention to introduce work study has been made known to everyone in the organization, it is not advisable to recruit work study staff or have them installed. If work study staff were installed without a clear statement of their function being made all kinds of misunderstanding could arise. It may take several months to recruit and train work study staff, during which time the others in the organization, e.g., the managers, engineers, supervisors, foremen, chargehands, trade union officials and shop stewards, can be provided training in work study.

The whole work study project should be carried out in such a way that it does not give any impression of being a "fault-finding mission". Everyone in the organization should be convinced of the necessity and benefits of work study.

Questions for Review
1. What are the possible reactions toward a work study program?
2. What is the importance of work study?
3. In an organization, how should work study be introduced?

11 SUMMARY AND REVIEW

Work study is concerned with studying the methods of working so that there will be more economical use of manpower, machinery, materials and money.

It involves method study to improve the method itself and work measurement to evaluate human effectiveness.

In its simplest form work study has been practiced since prehistoric times but has developed rapidly after the Industrial Revolution. Notable pioneers of work study were Robert Owen, Jean Perronet, F. W. Taylor, Frank and Lilian Gilbreth and Henry Gantt.

Method study involves the systematic recording and critical examination of existing or proposed work methods as a means of introducing more effective methods of work.

The most frequently used approach to method study is by means of the Method Improvement Master Sheet which consists of the following steps, wherein at every step the human factor is taken into account so that the commitment, cooperation and involvement of those affected by the new system are obtained:-

Select
Record
Examine
Develop
Install
Maintain

Recording is often carried out by means of Process Charts - Operation Process Charts, which provide an overall view, and, Flow Process Charts, which show operations and inspections as well as transportations, delays and storages.

Process Charts may be either Man-type or Material-type; they should normally follow either a person or the material, as mixing the two causes confusion.

Multiple Activity Charts allow the activities of different operators/units of equipment to be recorded at the same time to identify inefficiencies in their use, e.g., unnecessary waiting or delay by one or the other.

Two-handed Charts allow the same principle to be applied to the left and right hands - any under-utilization of either hand becomes evident when the activities are recorded simultaneously on paper.

Every element in the job has to be examined and questioned after all the facts are recorded. The questioning technique used here involves the following six questions:

What?
Where?
When?
Who?
How?
Why?

The activities which are challenged first are the "do" activities.

A better work method is developed as a result of the ideas that emerge at the challenging stage.

The improved work method has to be installed and maintained. Care should be taken to <u>explain</u> it clearly to those who are to implement the new work procedure and training should be provided for them. The checklist on introducing changes should be helpful in minimizing any resistance to the new work procedure.

ACTION GUIDELINES
[1] How can organizational effectiveness be improved through work study?
[2] If work study is not utilized in your organization, what is the main reason for it?
[3] In your organization, what are the main steps which should be taken to bring about work method improvement?

Questions for Review
1. What two main divisions is work study divided into?
2. Should work measurement be employed before work study, or, vice versa?
3. What is the reason?
4. Define method adjustment.
5. Give the six steps of work method improvement shown on the Method Improvement Master Sheet.
6. (a) Are all operations "do" activities?
    (b) Are operations and inspections only included in an operation process chart?
7. What is the reason "do" activities should be challenged first?
8. Draw and identify the five process charting symbols.
9. Name a chart which records simultaneously the activities of two or more persons.
10. Define the questioning technique. What are the six main questions employed in the questioning technique?
11. Why is it important to chart the new work method one recommends?
12. Why is it important to consult and involve employees when one develops a new work method?
13. What measures can be taken to ensure the successful adoption of a new work method by the employees who are affected?

12 APPENDICES

APPENDIX I

APPENDIX II

FLOW PROCESS CHART (MATERIAL)

PRESENT METHOD

JOB DESCRIPTION: Battery recharging

CHART BEGINS: Battery at reception rack   DATE: 4.9.81
CHART ENDS: Battery at delivery rack   CHARTED BY : D.L.

	SYMBOL	DESCRIPTION
	1 (transport)	Battery deposited at reception
	1 (delay)	Await for receipt work
	1 (operation)	Marked
	2 (transport)	Stacked
50M	2 (delay)	Await transporting
	3 (transport)	Placed on stillage truck
	4 (transport)	Moved to charge-shop
	3 (delay)	Await testing
	1 (inspection)	Volt test
	2 (inspection)	S.G. test
	2 (operation)	'Topped up' to the correct level
	5 (transport)	Placed on charging bench
	3 (operation)	Connected to test bar
	4-3 (combined)	Charged/tested
	5 (operation)	Disconnected
	4 (inspection)	Volt test individual cell
	6 (transport)	On stillage
35M	7 (transport)	To delivery bay
	8 (transport)	Placed out rack
	(storage)	Await collection

APPENDIX III

IMPROVED METHOD

Reach for U-bolt in bin 1	⇦	⇨	Reach for U-bolt in bin 5
Grasp bolt	○	○	Grasp bolt
Carry to central position	⇦	⇨	Carry to central position
Place bolt on jig	○	○	Place bolt on jig
Reach for casting in bin 2	⇦	⇨	Reach for casting in bin 4
Grasp casting	○	○	Grasp casting
Carry to central position	⇦	⇨	Carry to central position
Assemble casting onto bolt	○	○	Assemble casting onto bolt
Reach for 1st nut in bin 3	⇦	⇨	Reach for 1st nut in bin 3
Grasp nut	○	○	Grasp nut
Carry to central position	⇦	⇨	Carry to central position
Assemble nut onto bolt	○	○	Assemble nut onto bolt
Reach for 2nd nut in bin 3	⇦	⇨	Reach for 2nd nut in bin 3
Grasp nut	○	○	Grasp nut
Carry to central position	⇨	⇦	Carry to central position
Assemble nut onto bolt	○	○	Assemble nut onto bolt
Grasp whole assembly	○	○	Grasp whole assembly
Moves to finished product's box	⇦	⇨	Moves to finished product's box
Drop assembly to box	○	○	Drop assembly to box

IMPROVED LAYOUT

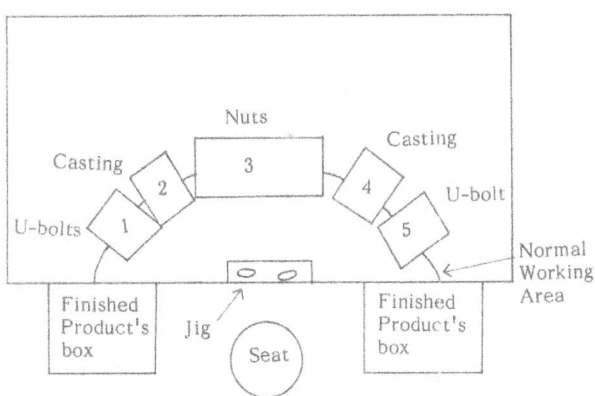

Sketch of a jig

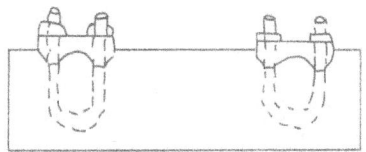

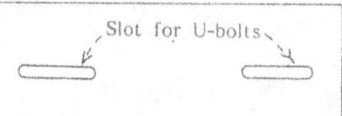

www.ingramcontent.com/pod-product-compliance
Lightning Source LLC
Chambersburg PA
CBHW081728170526
45167CB00009B/3738